Advanced Structured Materials

Volume 237

Common engineering materials are reaching their limits in many applications, and new developments are required to meet the increasing demands on engineering materials. The performance of materials can be improved by combining different materials to achieve better properties than with a single constituent, or by shaping the material or constituents into a specific structure. The interaction between material and structure can occur at different length scales, such as the micro, meso, or macro scale, and offers potential applications in very different fields.

This book series addresses the fundamental relationships between materials and their structure on overall properties (e.g., mechanical, thermal, chemical, electrical, or magnetic properties, etc.). Experimental data and procedures are presented, as well as methods for modeling structures and materials using numerical and analytical approaches. In addition, the series shows how these materials engineering and design processes are implemented and how new technologies can be used to optimize materials and processes.

Advanced Structured Materials is indexed in Google Scholar and Scopus.

P. M. Visakh
Editor

Engineering Applications of Polymer-Based Nanocomposites

Springer

Editor
P. M. Visakh
Natural Bioactive Materials Laboratory
Department of Bioengineering
Ege University
Bornova, Türkiye

ISSN 1869-8433 ISSN 1869-8441 (electronic)
Advanced Structured Materials
ISBN 978-3-031-93462-9 ISBN 978-3-031-93463-6 (eBook)
https://doi.org/10.1007/978-3-031-93463-6

This Springer imprint is published by the registered company Springer Nature Switzerland AG
The registered company address is: Gewerbestrasse 11, 6330 Cham, Switzerland

Contents

About the Editor

Dr. P. M. Visakh (M.Sc., M.Phil., Ph.D.) is a prolific editor with more than 50 books already published. Currently he is working in the Natural Bioactive Materials Laboratory, Department of Bioengineering, Ege University, Bornova/Izmir, Türkiye. He was working as Associate Professor in TUSUR University, Tomsk, Russia from 2017 till 2022. He did his Post-doctoral Research in Tomsk Polytechnic University, Tomsk, Russia (2014–2017). He obtained his Ph.D., M.Phil., and M.Sc. degrees from School of Chemical Sciences, Mahatma Gandhi University, Kottayam, Kerala, India. He edited 50 books from Scrivener (Wiley), Springer, Royal Society of Chemistry, Elsevier and more than 25 books in press (from Wiley, Springer, Royal Society of Chemistry, and Elsevier). He has been invited as a visiting researcher in Russia (2014–2022), Portugal (2013, 2014), Czech Republic (2012, 2013), Italy (2009, 2012), Argentina (2010), Sweden (2010–2012), Switzerland (2010), Spain (2011, 2012), Slovenia (2011), France (2011), Belgium (2012), and Austria (2012) for his research work. He visited 13 countries; He visited 15 universities in Europe. He published 23 publications, four reviews, and more than 55 book chapters. He has attended and presented more than 28 conferences, he has 2332 citations, and his h-index is 27. He acts as guest editor for four International Journals.

Chapter 1
Engineering Applications of Polymer-Based Nanocomposites—Challenges and Opportunities

Shiv Kumari Panda and P. M. Visakh

1.1 Polymeric Nanocomposites and Their Applications

Composites can be divided into three main groups: metal matrix composites, ceramic matrix composites, and polymer matrix composites. Polymer matrix composites or polymer nanocomposites (PNCs) are combinations of two or more materials. It has polymers as the main matrix and fillers as the dispersion phase (Oliveira and Beatrice 2019; Gulcihan Guzel Kaya 2022). Polymer matrix nanocomposites typically exhibit a phase boundary between the two components, forming an interfacial layer at this boundary. The interfacial region's characteristics, composition, and minute structure differ from those of both the polymer and the filler. A strong bond at the interface greatly enhances the overall properties of the nanocomposite (Thoams 2007). Two primary methods are used for this dispersion. One is chemical method which employs *in situ polymerization*, where nanofillers are first mixed in a monomer solution. The polymerization process then proceeds, allowing the nanoplatelets to swell in the monomer. As the polymer forms, it intercalates into the layered structure of the nanoplatelets. In essence, the polymer chains grow between the intercalated sheets, creating a uniform dispersion within the matrix. The scanning electron microscope (SEM) is a powerful imaging tool that uses a focused beam of electrons to produce highly magnified images of samples (Raja and Barron 2024). This analysis does not offer specific details regarding the filler's dispersion, distribution, or

S. K. Panda (✉)
Assistant professor, Department of Chemistry, Udayanath Autonomous College of Science & Technology, Adaspur, Cuttack, India
e-mail: shivkumaripanda@gmail.com

P. M. Visakh
Research Associate, Department of Bioengineering, Natural Bioactive Materials Laboratory, Ege University, Bornova/Izmir, Izmir, Turkey

P. M. Visakh (ed.), *Engineering Applications of Polymer-Based Nanocomposites*, Advanced Structured Materials 237, https://doi.org/10.1007/978-3-031-93463-6_1

interaction with the polymer matrix. However, it does provide insights into whether the filler is uniformly dispersed throughout a significant volume of the material. However, the bonding between fillers and the polymer matrix may not be effective always. Instead, they often exhibit a readiness to bond among themselves, leading to the formation of aggregates, known as filler-filler interactions. These interactions can weaken the properties of the composite material. In SEM grayscale images, these filler clusters or agglomerates are typically observed as white regions. One major challenge of TEM analysis is the preparation of extremely thin samples to allow sufficient electron transmission with minimal energy loss. TEM gives information about particle size, shape, distribution, crystallinity, crystallite size, structure, and orientation (Venkateshaiah et al. 2020). This microscopic information helps the researchers to understand the underlying mechanisms involved in materials. The scale of data obtained through AFM nanomechanics closely aligns with the resolution of computer simulations, offering a bridge between simulation results and the macroscopic properties of materials. These techniques have become invaluable for understanding the mechanical behavior of polymeric materials. The absorption of infrared radiation at various wavelengths can detect functional groups in organic and inorganic compounds. The process begins with the collection of an interferogram from the specimen by an interferometer. To generate the spectrum, the Fourier transform, a mathematical algorithm is applied to the interferogram. Modern FTIR spectrometers can digitize interferograms, perform the Fourier transform, and display the resulting spectrum. The sample absorbs the IR radiation if the radiation's frequency matches the vibrational frequency of a particular bond within the molecule. This absorption corresponds to the energy associated with functional groups, making IR spectroscopy a valuable tool for understanding chemical structures. However, chemical interactions between the polymer chains may occur for miscible components, resulting in changes to the spectra of the nanocomposite compared to those of the pure components. Shifts in wave numbers and the broadening of absorption bands in the spectrum are indications of these interactions (Jaleh and Fakhri 2016).

Differential scanning calorimetry (DSC) is a thermo-analytical method employed to measure the difference in heat flow between a sample and a reference, kept at identical temperatures, as their temperatures are increased during the experiment. In a physical transformation such as a phase change, the heat flow to the sample differs from that of a reference sample maintained at the same temperature. The direction of heat transfer determines whether the process is endothermic or exothermic. Metal-polymer composites function as metal-dielectric combinations, where collective oscillations of electron clouds, known as Surface Plasmon Resonance (SPR), occur at the interface between metal nanoparticles and the dielectric matrix. These SPR phenomena arise from the interaction between incident light and metal nanoparticles, contributing to the improved performance of solar cells (Muhammed Shameem et al. 2021). The composites offer the potential to enhance the dielectric constant and breakdown field simultaneously, resulting in substantial improvements in energy storage capacity (Riggsa et al. 2015). The graphene-polyethylene-dioxythiophene (G-PEDOT) structure significantly enhances the electrochemical performance of PEDOT-based supercapacitors. Polymer nanocomposites (PNCs) possess notable properties such as high adsorption capacity, along with catalytic and magnetic functionalities, which make them effective for dye removal. The incorporation of metals creates functional sites, enabling the design of various hybrid structures for this application. For example, metal-decorated polypyrrole nanocomposite anodes have been investigated for the electrochemical oxidation of Direct Red 80, achieving degradation rates ranging from 83% to 100%. This PVA/GO nanocomposite filter is a promising solution for treating indoor air contaminated with gaseous pollutants and holds potential for various industrial applications (Mohamed et al. 2023).

Nanobiocomposites can act as devices for sustained or targeted drug delivery due to their advantages, including biodegradability, biocompatibility, safety, environmental friendliness, and affordability. Recently, several nanobiocomposite-based drug delivery systems have been introduced commercially. Among biopolymers, starch-based materials such as alginate and chitosan are currently considered ideal for the synthesis of nanobiocomposites. In situ solution polymerization and emulsion polymerization are used for producing these nanobiocomposites (Rahim et al. 2020). Recent advancements emphasize the potential of thiophene-based green polymers and eco-friendly synthesized or functionalized fullerenes for applications in energy conversion, sensing, and diode devices. Utilizing green nanocomposites processed through environmentally friendly solution-based methods offers an effective route to achieve optimal results in this field (Carpenter et al. 2015). The addition of nanofillers such as layered silicate clays, cellulose nanofibers, metallic nanoparticles, and chitin nanofibrils into biopolymer greatly enhances thermal and dimensional stability, and this is due to the nanofillers' higher modulus and lower thermal expansion coefficient than the polymer matrix. Thus, the films will get greater stability against shrinkage and deformation during the processing of food packaging materials (Shankar and Rhim 2016).

1.2 Starch-Based Nanocomposite for Food Packaging Applications

Starch is the most important polysaccharide, and it is the most dominant storage form of carbohydrate in the plants. The chief sources of starch are grains such as corn, wheat, rice, and vegetables such as potato, sweet potato, cassava, tapioca, in fruits such as apple and banana (Apriyanto et al. 2022). Starch is the versatile form of biomaterial due to its abundance, readily bioavailability, low cost, non-toxic property, and biodegradability used in many food and non-food industries (Madhumitha et al. 2018). Waxy starches in paste form are advantageous than amylose due to its high stability toward retrogradation; the size of starch granules directly depends on the compactness of amylopectin. Although starch remains insoluble in aqueous medium at room temperature, when the temperature increases, an irreversible swelling process occurs, namely gelatinization, which leads to the dissolution of starch by degradation of amylose and amylopectin. The solubility of the film is an important factor in the use of these materials as food packaging. The highly soluble films are frequently appealing for the dried item that requires hydration before consumption. Water solubility is a significant attribute of biodegradable films and is a measure of water-soluble content in a film. The standard value for film solubility is based on its purpose or intended use for preparing food products as they are soluble in the consumer's mouth (Singh et al. 2018). The gelatinization processing involves granular swelling, native crystalline melting, and molecular solubilization (Jiang et al. 2020). Starch granules are made of alternating layers of the amorphous and crystalline lamellae with thicknesses between 100 and 400 nm. Testing and experiments are going on to characterize the property of the starches extracted from unconventional sources or from new different sources. Normally starch from maize, potato, wheat, rice, sorghum is used in large amount in the field of pharmacy (Phukan and Nongkhlaw 2022).

The science of nanotechnology has been attracting the attention of researchers and scientists since from past decades, which actually deals with the formation and designing of nanosized particles ranging from 1 to 100 nm (Hasanin et al. 2022). In recent days, nanocomposites have gained much attention over traditional composite materials and are widely used in food packaging, biomedical applications, electronics, energy storage, optics, the automotive industry, bio-sorbants for environmental remediation, textiles, and many other applications. Researchers are exploring starch modification techniques to improve its properties and develop novel composites. Starch can be modified into nanoparticles and can also undergo various physical (milling, blending with other polymers, extrusion, plasticizers, etc.) and chemical (substitution, graft co-polymerization, cross-linking, oxidation, etherification, esterification, dual modification, etc.) modifications to produce materials with novel properties (Ates et al. 2020). In the preparation of starch-based bionanocomposite, starch is used as an organic matrix to which various nanoreinforcements are added to improve its properties, such as mechanical and thermal properties, oxygen barrier, and biodegradation rates. The successful bonding between the polymers is achieved by homogenous mixing, thus improves the material properties and also influences the crystalline and molecular network of matrix. The nature of starch and its cohesive structure is influenced by processing methods, environmental conditions, type of plasticizer, thickness, and other factors that determine the properties of the film (García-Guzmán et al. 2022).

Avella et al. (2005) discovered that the starch–clay nanocomposite could be used for the food packaging purpose. The use of the nanoparticle in the different starch-based clays was applied to produce the starch-based film for packaging. Use of such materials for preparing films increases the properties of the packaging films and also tensile strength and modulus. The food package should hinder gain or loss of moisture, prevent microbial contamination, and act as a barrier against permeation of water vapor, oxygen, carbon dioxide, and other volatile compounds such as flavors and taints in addition to the basic properties of packaging materials such as mechanical, optical, and thermal properties (Rhim et al. 2013).

1.3 Bionanocomposites for Food Packaging Applications

Food packaging materials need to have a certain fundamental quality. These materials have the ability to preserve food safety along with quality. These have the ability to block negative elements, including chemical pollutants, spoiling microbes, moisture, oxygen, light, and outside forces, among others. Components of food packaging serve as a barrier to prevent the infiltration of volatile substances (taints and tastes), water vapor, oxygen, and carbon dioxide. These substances have favorable mechanical, thermal, and optical characteristics (Mohanty and Swain 2017). The increased surface area and big aspect ratio of nanofillers are responsible for these enhanced qualities. These substances are referred to as bionanocomposites. According to some reports, bionanocomposites made for packing applications offer better antibacterial, mechanical, thermal, rheological, and barrier qualities. These materials can be referred to as smart packaging materials (Rhim and Ng 2007). Nanocomposites have been synthesized, and their properties have been investigated in order to preserve or even enhance the food's quality, active packaging involves a series of interactions between the containers and the food. The filler inserted in the polymer matrix at the nanoscale might be either organic (like nanocellulose) or inorganic (like silica clays). Furthermore, the filler produces more extensive interfacial interactions with the organic matrix and improves performance compared to the component of polymers. Because they are not biodegradable, the polymers frequently utilized to create polymer nanocomposites are less environmentally benign (Puiggalí and Katsarava 2017). Layered nanocomposites can therefore be flocculated, exfoliated, and intercalated (or phase separated). As with hydroxyapatite (HAp), it should be taken into account that the inorganic part of hybrid nanocomposite may have a natural origin. In this instance, this could potentially improve the ultimate biocompatibility.

Metal ions are reduced to metal atoms in the first stage, and the reduced metal atoms are subjected to nucleation. The second stage is characterized by the simultaneous rise in thermodynamic stability and the coalescence of small nearby NPs into bigger particles. NPs are given their ultimate shape at the final stage, when the process is terminated (Dikshit et al. 2021). In order to facilitate polymer formation between the intercalated sheets, this approach disperses the nanoparticle in a liquid monomer or a monomer solution. Heat, radiation, the diffusion of an appropriate initiator, an organic catalyst, or both can be used to carry out polymerization (Pande and Sanklecha 2017).

1.4 Polymer-Based Nanocomposites for Water Remediation: From Adsorption to Photodegradation

Wastes from agriculture setups, dump yards, and untreated sewage are also resulting in water pollution directly or indirectly. According to a WHO report on 'Drinking Water' in 2017, nearly 50 % of the world's population will face water stress by the year 2025 and 21% will have no sanitation service in less developed nations (WHO 2017). In poorly developed nations, water-borne diseases like cholera, diarrhea, and hepatitis are more prevalent (Hatami 2013). The most frequently utilized polymers include polyethylene, polyacrylamide, poly(vinylidene fluoride), polysulfone, cellulose acetate, polyethersulfone, polycarbonates, polyacrylonitrile, and polypropylene. Nanotechnology is also rapidly growing and attracting more scientific interest due to the functional characteristics of nanomaterials (Lu et al. 2016). Nanomaterials are materials smaller than 100 nm in at least one of the dimensions and because of their extremely small size and high surface area, nanomaterials possess large adsorption capacity and reactivity (Baig et al. 2021). Moreover, all the nanomaterials have well-defined and controllable morphologies of proper size and porosity and hence can be efficient platforms for water remediation (Li et al. 2015). Polymer nanocomposites are materials made by incorporating nanomaterials or nanoscale fillers into the polymer matrix. These fillers are usually in the form of NPs, nanofibers, or nanolayers, which are dispersed within the polymer to enhance or modify its properties (Carpenter et al. 2015). These nanocomposites can be used as effective adsorbents, membranes, and photo-reactive agents for water remediation (Adeleye et al. 2016; Otrisal et al. 2018).

Different factors could affect the adsorption and photodegradation mechanisms of polymer nanocomposites to be applied in water treatment. The removal of various pollutants and contaminants from water with polymer nanocomposites is based on different phenomena such as adsorption, photocatalysis, oxidation, co-precipitation, and filtration (Chenab et al. 2020; Santhosh et al. 2016). This adsorption approach can be effectively used as a water purification system (Santhosh et al. 2016; Jiang et al. 2018). On the other hand, TiO_2, CNTs, and graphene-based polymer nanocomposites are used in photocatalytic degradation, which is a more advanced technology for the breakdown of trace contaminants, organic pollutants, and microorganisms (Zaidi et al. 2019). These fillers which can include NPs, nanotubes, or nanoclays, are dispersed throughout the polymer. Various techniques like direct compounding, solution casting, intercalation, in situ polymerization, sol-gel synthesis, electro-spinning, and electrochemical methods have been employed for creating polymer nanocomposites (Blanco et al. 2019). Polymer nanocomposites can function as adsorbents, coagulants, membranes, and photocatalysts to target specific contaminants with improved efficiency and selectivity (Kundu 2022). Biodegradable natural polymers present significant benefits as prospective matrix components in nanocomposites. Common natural polymers, such as polysaccharides, proteins, and lipids, possess the ability to create biocompatible nanocomposites (Imamura et al. 2013).

The incorporation of a polymer matrix significantly enhances the processability of non-polymeric components within the composite, allowing for easier fabrication and manipulation. Polymers are often selected as host materials due to their favorable processability, making them ideal candidates for creating nanocomposites. The main mechanisms for water treatment by these materials include adsorption, chemical degradation, and photodegradation. TiO_2 is the most outstanding photocatalyst due to its strong photocatalytic properties, photostability, and cost-effectiveness (Guesh et al. 2016; Umar et al. 2019). To generate charge separation inside the particles,

TiO_2 significant bandgap energy (3.2 eV) requires UV irradiation. TiO_2 produces reactive oxygen species (ROS) in response to UV irradiation, which can destroy pollutants in a short reaction time (Moon et al. 2014; Babele et al. 2021). Their benefits for wastewater treatment stem from their high adsorption capacity, fast kinetics, aromatic selectivity, and large surface area (Saifuddin et al. 2013; Vairavapandian et al. 2008; Sadegh et al. 2016). Graphene and GO possess efficient adsorption for a variety of contaminants, including dichlorobenzene, ethylbenzene, Zn^{2+}, Pb^{2+}, Cu^{2+}, and Cd^{2+}, and dyes (Ren et al. 2011; Peng et al. 2003; Lu et al. 2008; Li et al. 2003; Moradi et al. 2017). The most common radical species generated during the photocatalytic reaction is the hydroxyl (OH^-) radical when the organic contaminant is in contact with the photocatalytic material (Pandey et al. 2020; Naushad et al. 2021). The photodegradation approach is widely popular for the degradation of organic dyes and heavy metals in photocatalytic reactions. Natural polymer-based nanocomposites have gained recognition as effective materials due to their environment-friendly nature. In this study, the starch-ZnO nanocomposite comprised a suspension of cross-linked starch and a gel of ZnO (Kamari and Shahbazi 2021). This was utilized as a multifunctional adsorbent to extract toxic Pb^{2+} ions from the aqueous environment. The resulting starch-ZnO nanocomposite exhibited rapid adsorption equilibrium for Pb^{2+} within 120 minutes. In another study, a nanocomposite membrane was developed by combining $Fe_3O_4@SiO_2$–chitosan (Zhang et al. 2014a).

1.5 Thermoplastic Starch-Based Composites

The starch obtained from various sources has varying levels of amylose and amylopectin. The amylose content varies within the same botanical verity depending on geographic aspects and climatic conditions. It facilitates the synthesis of TPS films with different mechanical properties (Muscat et al. 2012). Muscat et al. studied the outcome of using starch with different amylose content on TPS properties (Wunderlich 2011). The tensile strength and elastic modulus values were higher, and elongation at break was lower for films with a higher amount of amylose. Starch is a natural polymer having composite levels of organization that impose upon thermoplastic distortion. The polymers that can flow as a result of heating beyond their melting temperature are known as thermoplastics. The main reasons underlying the plastic deformation of the polymers are its increased molar mass, enmeshment, interaction, and chain branches (Perez-Pacheco et al. 2016). Thermoplastic starch (TPS) is obtained due to alteration in the structure of the starch particle. This alteration is caused by processing the starch by applying temperature and shear force, along with a small quantity of water and plasticizing agents, which do not evaporate quickly during preparation (Zhang et al. 2014b). Biodegradable polymers such as polyvinyl alcohol, polyoxyethylene, and polycaprolactone are incorporated with TPS to decrease TPS limitation, closely related to the hydrophilicity of TPS. After the starch film formation, the incorporated polymeric compound will even out the amorphous configuration by forming hydrogen bonds with the starch. It prevents the recrystallization of the starch from forming V-type crystals through the process of retrogradation. Good adhesion is extremely wanted for sealing purposes of food packaging film. In the absence of sufficient adhesion, food may turn out to be contaminated with food pathogens. Surface tension also forms a vital factor for wettability. The high value of surface tension leads to the high-quality distribution of a liquid over the surface (Zhou et al. 2009).

The surface properties of TPS are affected by the ratio of amylose to amylopectin, pulsed light, chemical modifications, and plasticizers. The incorporation of sodium benzoate as a photosensitive agent in TPS/PVA under UV radiation results in improved hydrophobicity, tensile strength, and Young's modulus of the composite blend (Gutiérrez et al. 2018). This was due to the enhancement in the cross-linking density of the macromolecular chains over the surface of the composite films (Liu et al. 2010). The starch granules are degraded as a result of thermal and mechanical action inside the extruding machine. The amylopectin molecules are affected to a greater extent than amylose molecules. The fragmentation and degradation of the polymer chain occur selectively near the center of the starch molecule (Gomez and Aguilera 1984). The mechanical and rheological properties of the starch are influenced by the variation in the molecular weight of starch and the molecular weight distribution of starch. The degradation in the molecular weight of starch throughout the thermoplastic starch preparation is influenced by the temperature applied, the amount of plasticizer used, and the magnitude of specific mechanical energy used. The resulting product characteristics were used to study the influence of water on the extrusion of maize starch (Azevedo et al. 2019). Starch granules occur in various shapes and sizes based on the source from which it was obtained. The starch granules may be oval, elliptical, circular, or polygonal (Bulatović et al. 2019).

In the study performed by Bulatovic et al. (Zain et al. 2019), the SEM examination showed that the surface of the composite of polycaprolactone (PCL) with 30 weight % TPS was smooth with a small surface roughness that was caused due to imperfectly plasticized starch. The smooth surfaces are characteristic of the PCL matrix that degrades at a slower rate, and sharp holes are characteristic of the thermoplastic starch phase that degrades faster. The thermal properties of the TPS composites are studied using thermogravimetric analysis (TGA) and differential scanning calorimetry (DSC). The TGA is used to infer the quantity and rate of variation in the weight of a substance with respect to time and temperature in a controlled environment (Iguchi et al. 2000). The DSC is used to study the differences in the thermal properties of composites.

1.6 Preparation, Characterization, and Antibacterial Activity of Biodegradable Bacterial Cellulose Nanocomposites

The production of the nanofibrillated cellulose by *Acetobacter xylinum* involved the conversion of the polymerized glucose units into $\beta(1 \rightarrow 4)$ glucan chains (Klemm et al. 2005). The bacterial cellulose can grow in static culture conditions; it will also produce thick pellicles with unique properties such as improved mechanical properties, high crystallinity, purity, and a homogeneous thick gel sheet-like structure (Donini et al. 2010). Even though plant-based cellulose has the properties mentioned above, they are not pure cellulose; it contains lignin and hemicellulose. Various chemical and physical treatments do the isolation of the cellulose nanofibers or cellulose nanoparticle. The microbes that were producing bacterial cellulose are reported from the species of Agrobacterium, Gluconacetobacter, and Sarcina (Güzel 2024). The bacterium *Gluconobacter* can able to convert the ethanol to acetic acid (Rahim et al. 2020 Oct). An early report mentioned that some fungi, seaweed, and non-pathogenic bacteria such as *K Xylinus* could also produce bacterial cellulose (Klemm et al. 2001). The important property of *K Xylinus* is that it can produce extracellular cellulose with a variety of thicknesses by monitoring oxygenation. The major advantages of bacterial cellulose are non-toxic, biocompatible, and biodegradable (Kim et al. 2007). The reactor-based production of bacterial cellulose is not a commonly practicing procedure due to the lack of feasible technology for large-scale cellulose synthesis from bacterial. Similarly, the formation of the negative cellulose mutants in the reactor is also reported as an issue in bacterial cellulose production. The common reactors preferred are static and agitated bacterial cellulose culture, rotary biofilm contactor, spin filler-attached bioreactors, rotary disk reactors, and silicon membrane-based reactors (The Nano-sized width of the bacterial cellulose is identified by Transmission Electron Microscope [79] n.d.).

Bacterial nanocellulose morphological aspect study analysis and visibility of the clear structure of the bacterial cellulose pellicles are possible with polarized optical microscopy. The naturally occurring cellulose can classify into two, bacterial cellulose and plant cellulose. Plant cellulose is mesh-like tough cellulose fibers, which can provide the plant with a basic plant structure (Sani and Dahman 2009). However, bacterial cellulose does not contain hemicellulose and lignin, which will be in the purest form of cellulose. Bacterial cellulose can be produced/molded-in sheet-shaped and has significant mechanical properties. Bacterial cellulose is macrostructured fibrils, which stacks in the form of bundles, and then this makes it to ribbon form, and it is known as pellicle (Steinbuhel 2001). Oxygen is an essential factor for the growth of cellulose-producing microorganisms. Nanocomposites are synthesized by two methods; they are ex situ and *in situ* methods. In the case of *in situ*-based synthesis, the nanomaterial is added along with the bacterial cellulose/polymer (Qiao et al. 2015). However, in the ex situ method, bacterial cellulose is impregnated with reinforcement materials (Ul-Islam et al. 2012).

The nanomaterials of metal and metal oxides of Ag, Au, TiO_2, and carbonaceous nanomaterials (carbon nanotubes, graphite) are commonly used to synthesize the bacterial cellulose nanocomposites (Kim et al. 2013). The bacterial cellulose nanocomposites are mainly used in the medical, paper, textile, leather, and food industry, due to its sustainability, structure, and biocompatibility (Jebali et al. 2013). Moreover, for the medical industry, bacterial cellulose-based composites are used (Falcão et al. 2008). The bacterial cellulose nanocomposites are capable of improving the antibacterial property, and it can be used for tissue engineering and drug delivery (Barud et al. 2011). In medicine, a bacterial cellulose-based wound dressing with antibacterial property is gaining attention nowadays (Vitta et al. 2010). The study reported that the bacterial cellulose with a porous network structured in dressing material could deliver medicine/antibiotics to the wound and create a barrier against external infection (Gao et al. 2014). In unhealed and absolute wounds, the bacterial cellulose-based nanocomposite dressing bandages can reduce bacterial enzyme activity and prevent an anaerobic environment. Thereby reduce the infection and hence heal the chronic wound (Raquel et al. 2019; Wippermann et al. 2009). The first and foremost bacterial cellulose-based wound dressing product available on the market was biofill (Fibrocel Produtos Biotechnologies LTDA (Fibrocel), Brazil). The main properties of the biofill are its thin structure and 85% water content. Hence, it can be used to treat ulcers and burns as a substitute for skin/skin grafts. The main advantage of biofills can fast relief pain and infection (Fontana et al. 1990). Wu and Lia (2008) experiments reported that the bacterial cellulose was used as an immobilizing material for glucoamylase. Immobilization of the glucoamylase enhanced stability at low pH and different temperatures. The bacterial cellulose has been used for bone grafts and bone-tissue engineering. Bone is made up of organic materials such as collagen/noncollagenous proteins and inorganic calcium hydroxyapatite. In a study, calcium chloride ($CaCl_2$) and sodium carbonate (Na_2CO_3) were added to the bacterial culture medium for promoting calcium carbonate deposition in bacterial cellulose.

1.7 Different Types of Biodegradation Analysis for Nanocomposites

Renewable resources or natural sources are mainly used in nanocomposites, leading to increasing biodegradability (Rutkowska et al. 1998). Natural sources for making the nanocomposites achieve desirable properties such as biodegradability, improved thermal, optical, and mechanical properties. Usage of organic nanoparticles (cellulose, chitosan, and starch) increases the biodegradability compared to inorganic nanoparticles (silver, titanium, and copper) in the nanocomposite. The same study can also be carried out in laboratory conditions with the help of a reactor filled with sand and seawater (Rutkowska et al. 2002). The reactor should also be kept in a closed condition to avoid excessive drying out and vapor release. The sand and seawater should be collected from the littoral zone to contain an increased organic matter. After creating the preferred condition, the biodegradation study sample was placed in a framework of specific parameters and buried in the sandy bed. Then, the sample weights were taken at regular intervals of time (Weber et al. 2006). Bionanocomposite has the advantage of faster degradation than normal plastics, thereby reducing the accumulation frequency. It is thus important to study the soil biodegradability of the bionanocomposite produced. Different methods are carried out so far, which can be chosen based upon the material produced and equipment available (Raghavan 2006). Polyvinyl alcohol (PVA) and polycaprolactone (PCL) are the two synthetic polymers widely used to produce nanocomposites. PVA composites have a greater degradation percentage as compared to PCL composites. The degradation rate of PCL can be improved by incorporating bionanoparticles than inorganic nanoparticles. Bionanoparticles such as cellulose, starch, and chitosan will increase the biodegradation rate as they are more prone to microbial attacks due to their more straightforward structure (Sarwar et al. 2018). Nanocellulose can be produced in different forms such as crystals, fiber, acylated CNF and are reported to increase biodegradability.

Microbial Oxidative Degradation Analyzer (MODA) can be used to find out the degradation rate of bionanocomposite based upon the amount of carbon dioxide produced. The analysis is based on the standard procedure as per ISO 14855-2 (Kunioka et al. 2007). The analyzer contains a reaction vessel which is maintained at uniform temperature and humidity throughout the reaction. These are controlled heater and by-passing humid air inside the reactor. The reactor can be filled with about 400g sea sand mixed with activated compost. 10g of the composite sample can also be mixed with the sand, and then composting can be done. Mass loss is a major physical change during composting. Mass loss occurs due to the conversion of organic matter in the polymer composite to carbon dioxide and water, which gets escaped and evaporated into the atmosphere. The mass loss will further reduce the mechanical and thermal properties of the polymer. Thus, polymer degradation can be analyzed by tracking the mass loss and loss of mechanical properties during the study period (Panthapulakkal and Sain 2012; Eckel et al. 2024). Microbial degradation of bionanocomposite can also be analyzed by weight loss due to organic matter utilization for microbial growth. A specific amount of composite can be kept inside a microvial filled with a fixed amount of bacterial culture. The vial can then be incubated at a specific temperature for about two weeks. Vials should be prepared in sufficient number to record the weight loss intermittently during the study period (Ahmada and Sultanaa 2017; Mndlovu et al. 2024). Microbes break down the polymers by producing enzymes, and these enzymes can be produced externally and added to polymers for faster degradation compared to microbial degradation. Testing can be done by placing the specimen in a medium with the required amount of enzymes. A CO_2 evolution test can be done to find out the degradation rate. Percentage biodegradation can also be determined by analyzing the percentage weight loss of strength loss of polymer composite immersed in the enzyme (Spiridon et al. 2008; Shalem et al. 2024).

1.8 Exploring the Role of Magnetic Nanoparticles in Cancer Imaging and Diagnostic Advancements

People have been exploring magical ways to reduce the pain of illness and injury for hundreds of years. Numerous scholars believe that accomplishing this objective may depend on the application of nanotechnology in medicine. These applications use molecular level nanostructures and nanodevices to generate, monitor, control, repair, and safeguard entire biological human systems. Nanotechnology has the potential to revolutionize medical research and open up new fields in human development, and there is a strong road for using nanotechnology in medicinal, preventative, and diagnostic applications. This technology offers tools to investigate tissue structure at the nanoscale, aiding in understanding biological processes between atomic and cellular levels (Mura et al. 2013). Some nanoparticles have possible applications in diagnostics, imaging, and therapy/treatment such as targeted medicinal products, pharmaceutical products, biomedical implants, and tissue engineering. Biomedicine also now stands at another threshold with the full potential of nanobiotechnology, with the use of artificial intelligence for predictive designs, assays and data results and treatments. In this chapter, different types of nanoparticles will be discussed used in the medicine and they will be explained applications of that nanoparticles in diagnostics and imaging (Coskun 2016).

In recent years, numerous studies have been conducted to investigate the clinical applications of nanotechnology for the detection, diagnosis and treatment of cancer and other diseases, and efforts have been increasing to develop a variety of organic or inorganic nanoparticles, which can be categorized as different nanoscale materials such as quantum dots, gold nanoparticles, iron oxide nanoparticles, upconversion nanophosphors, metal-organic lattices and zeolites, and purely inorganic or purely organic materials such as liposomes (Gao et al. 2004; Huang et al. 2006). Nanoparticles are also used in various cancer-fighting therapies and applications such as chemotherapy, photodynamic therapy, neutron capture therapy, thermal therapy, and magnetic therapy (Bardhan et al. 2011; He et al. 2015; Petros and DeSimone 2010). In the 1990s, clinical molecular imaging emerged as a promising and minimally invasive in vivo tool to visualize cellular functions and monitor cells with the perception that the underlying causes of most modern diseases could only be understood through increasingly specific cellular or subcellular studies. In recent years, advancements in biomedical imaging technologies have significantly expanded the capabilities of existing modalities. The nanoparticles are small enough to get past the mucosal barrier and reach the cancer target, which is the cytoplasm or nucleus of the cell. Recent studies have shown that various functional nanoparticles can be used as drugs or targeting moieties for effective and site-specific delivery when covalently bound to a range of biological molecules, such as peptides, proteins, nucleic acids, or small molecule ligands (Henglein 1989). MRI is a noninvasive medical imaging technique, CT, SPECT, and PET. It avoids the use of ionizing electromagnetic radiation and the associated hazards; the energy emitted can interact and disrupt chemical bonds within living tissue (Reisz et al. 2014) that physicians use to diagnosis and whole medical treatment (Tran and Webster 2010). MRI is related to the powerful magnetic field, radiofrequency (RF), a computer to produce detailed pictures of organs, soft tissues, bone, and virtually all other internal body structures. All hydrogen atoms have a nuclear spin, which is related to a magnetic dipole moment, and are thus detectable by MRI. MRI scanners consist of two powerful magnets. Also, the human body is seventy percent water molecules, which consist of hydrogen and oxygen atoms. At each atom, center has a proton, which operates as a magnet and is sensitive to any magnetic field. As Rosen et al. pointed out, targeting techniques typically fall into one of two categories: active targeting, which uses ligands conjugated to the surface of SPIONs to identify specific surface landmarks in cancerous tissue, or passive targeting, which depends on physiological differences between cancerous and normal tissues.

These magnetic nanoparticles can be administered intravenously to produce effective in vivo transfection efficiency. The therapeutic genes are released by enzymatic cleavage, electrostatic interactions, or matrix degradation when the nanoparticles are kept at a target spot under the influence of high gradient electromagnetic fields (Bi et al. 2020).

1.9 Advancements in Green Nanocomposites: A Comprehensive Review on Cellulose-Based Materials in Biocomposites and Bionanocomposites

Cellulose is the most common and fundamental structural element in plant fiber. Cellulose contains a linear change of D–anhydroglucopyranose which is then combined with b–1–4 glycosidic (Thanapornsin et al. 2024). This structural combination results in the formation of the microfibrils. This structure has a compatibility of both crystalline and amorphous structures. The main reason behind the adaptability of this kind of structural behavior is because of its fibrous, tough, and hydrophilic nature (Salama 2023). Nanocomposite materials are formed with nanoscale dimensions and have unique properties which makes them attractive for their significant use in any industry. Most precisely, the polymer matrix nanocomposite (PMNC) is the most common one with its effective barrier, optical characteristics, and mechanical properties (Vishnupriya et al. 2024). These all properties make them individual elements for the diverse application. When it comes to the industrial point of view, the nanocomposite material has played an important role in the automotive and packaging industry with their lightweight and cost-effective solutions. The term bionanocomposite states that the material is made up of biological components and more diligently with polymers and inorganic solids. Compared to the traditional or the conventional nanocomposite material, the bionanocomposite material has some unique characteristics such as biodegradability, biocompatibility, and unique functionality. Some of the research works in the field of nanobiocomposites, green composites, and biohybrids have gained attention for the convergence of nanotechnology and biology (Kumar and Kumar 2022). Cellulose, which is the main component of plant cell walls, possesses a molecular structure that highlights its unique qualities (Verma et al. 2024). Cellulose is composed of linear chains of glucose units, which exhibit both crystalline and amorphous regions. These regions collectively contribute to cellulose's exceptional tensile strength, natural capacity to degrade, and remarkable stiffness.

The hydrogen bond arrangement of cellulose contributes to its thermal qualities, enhancing its appeal. Cellulose-based biocomposites exhibit exceptional resistance to heat degradation, making them highly suitable for applications that require strong temperature resistance. This provides opportunities in various industries, such as automotive manufacture, construction, and aerospace, where materials need to endure different heat conditions while maintaining their structural integrity. The process of creating cellulose-based biocomposites involves a dynamic interaction between renewable resources found in nature and advanced processing processes. This section explores the fascinating processes that convert cellulose into adaptable and environmentally friendly wonders, ranging from the extraction of cellulose to the complex techniques used for composite synthesis (Das et al. 2023). Cellulose-based biocomposites are at the forefront of the packaging industry, leading a shift towards materials that are both lightweight and environmentally benign. Envision packaging options that not only safeguard products but also actively contribute to a sustainable future. When combined with other polymers, cellulose creates packaging materials that redefine industry norms and promote environmental consciousness (Al Mahmud 2023). The partnership between material scientists and nanotechnology experts serves as the driving force for surpassing current constraints and advancing cellulose-based biocomposites into unexplored domains (Ruz-Cruz et al. 2022). The exploration of new horizons for cellulose-based biocomposites beckons researchers, extending the journey beyond present uses. The combined contributions of material scientists, engineers, and environmental experts play a crucial role in expanding knowledge and finding innovative uses that transform the field of sustainable, high-performance materials. Explore the fascinating realm of nanocellulose, a remarkable substance obtained from cellulose, where cellulose nanocrystals (CNC) and cellulose nanofibrils (CNF) play a prominent role. Imagine CNC as the mighty Hercules of the nanoworld, with extraordinary mechanical strength and impressive aspect ratios that surpass conventional boundaries (Lv et al. 2024).

Nanocellulose, with its ability to regenerate and sustainably meet growing world needs, can be likened to a phoenix that emerges from destruction. The low toxicity of nanocellulose ensures that innovation does not result in any harmful consequences. Within the complex realm of material science, nanocellulose possesses characteristics that not only serve as constituents of bionanocomposites, but also act as advocates for a sustainable future (Agbakoba et al. 2023). The manufacturing of cellulose-based bionanocomposites involves the complex task of incorporating nanocellulose into polymer matrices, a process that extends beyond simple blending. Envision a realm where the amalgamation of nanocellulose and polymer transcends into an artistic endeavor, molding the attributes of the bionanocomposite in a manner that redefines the realm of potential (Hellweg et al. 2023).

1.10 Significance of Life Cycle Assessment of Clean Wastewater Treatment Technologies for Achieving Sustainability

Life cycle assessment (LCA) provides a thorough procedure for assessing WWTPs' environmental performance overall (Babuji et al. 2023). With the use of life cycle assessment, we can assess the environmental costs and advantages related to the abstraction of raw materials, building, operation, maintenance, and eventually decommissioning of wastewater treatment plants. Through life cycle assessment, we can determine possible hotspots, comprehend trade-offs, and create more environmentally friendly wastewater treatment technologies. Hotspots are those points in the life cycle that have substantial negative influence on the environment. Diseases caused by microbes, heavy metals, chemical pollutants, pesticides, medications, and novel toxins are all potential water contaminants. Chemical contaminants, such as common household items, agricultural runoff, and industrial waste, pose a threat to human health because they contaminate water supplies. Ongoing research is being conducted to determine the effects of new pesticides, medications, and emerging pollutants on human health (Shen et al. 2012). The characteristics of non-point source (NPS) pollution differ from those of point source pollution, including the fact that it occurs at random intervals, is caused by complicated mechanisms and processes, has unknown discharge pathways and volumes, varies in pollutant loads throughout space and time, and is difficult to monitor, simulate, and regulate. Because of these features, NPS pollution is difficult to detect, regulate, and cure (Kamali et al. 2019). Based on the findings of the inventory analysis, the next step is to conduct a life cycle impact assessment (LCIA) to identify the environmental consequences across the stages of the life cycle that were established in Stage I. Possible environmental impact assessment metrics, such as GWP, are generated at the LCIA stage. Interpretation of LCI and LCIA data is the last step in an LCA research (Hijrah et al. 2023). Considering the gate-to-gate phase is crucial since the operational stage is a major contributor to the environmental burden. Because of the high stakes involved in managing a wastewater treatment plant—which might have far-reaching consequences for the environment if not addressed—this study zeroes emphasis on that part of the process [134]. PVDF membranes have also shown promise in water treatment applications, particularly due to their excellent chemical resistance and mechanical strength.

References

Adeleye AS, Conway JR, Garner K, Huang Y, Su Y, Keller AA (2016) Engineered nanomaterials for water treatment and remediation: costs, benefits, and applicability. Chem Eng J 286:640–662

Agbakoba VC et al (2023) Preparation of cellulose nanocrystal (CNCs) reinforced polylactic acid (PLA) bionanocomposites filaments using biobased additives for 3D printing applications. Nanoscale Adv 5(17):4447–4463

Ahmada N, Sultanaa S, Azamb A, Sabira S Khana MZ (2017) Novel bio-nanocomposites materials for enhanced biodegradability and photocatalytic activity. New J Chem 20:36:13:1–28
Al Mahmud MZ (2023) Exploring the versatile applications of biocomposites in the medical field. Bioprinting 36:e00319
Apriyanto A, Compart J, Fettke J (2022) A review of starch, a unique biopolymer–structure, metabolism and in planta modifications. Plant Sci 318:111223
Ates B, Koytepe S, Ulu A, Gurses C, Thakur VK (2020) Chemistry, structures, and advanced applications of nanocomposites from biorenewable resources. Chem Rev 120(17):9304–9362
Avella M, De Vlieger JJ, Errico ME, Fischer S, Vacca P, Volpe MG (2005) Biodegradable starch/clay nanocomposite films for food packaging applications. Food Chem 93(3):467–474
Azevedo VM, Carvalho RA, Borges SV, Claro PIC, Hasegawa FK, Yoshida MI, Marconcini JM (2019) Thermoplastic starch/whey protein isolate/rosemary essential oil nanocomposites obtained by extrusion process: antioxidant polymers. J Appl Polym Sci 136(23):47619
Babele PK, Verma MK, Bhatia RK (2021) Carbon nanotubes: a review on risks assessment, mechanism of toxicity and future directives to prevent health implication. Biocell 45(2):267
Babuji P et al (2023) Human health risks due to exposure to water pollution: a review. Water 15(14):2532
Baig N, Kammakakam I, Falath W (2021) Nanomaterials: a review of synthesis methods, properties, recent progress, and challenges. Mater Adv 2(6):1821–1871
Bardhan R, Lal S, Joshi A, Halas NJ (2011) Theranostic nanoshells: from probe design to imaging and treatment of cancer. Acc Chem Res 44(10):936–946
Barud HS, Regiani T, Marques RFC, Lustri WR, Messaddeq Y, Ribeiro SJL (2011) Antimicrobial bacterial cellulose-silver nanoparticles composite membranes. J Nanomater 2011:1–8
Bi Q, Song X, Hu A, Luo T, Jin R, Ai H, Nie Y (2020) Magnetofection: magic magnetic nanoparticles for efficient gene delivery. Chin Chem Lett 31(12):3041–3046
Blanco M, Monteserín C, Angulo A, Pérez-Márquez A, Maudes J, Murillo N, Aranzabe E, Ruiz-Rubio L, Vilas JL (2019) TiO2-doped electrospun nanofibrous membrane for photocatalytic water treatment. Polymers 11(5):747. (40). Shah V, Bhaliya J, Patel GM, Deshmukh K (2022) Advances in polymeric nanocomposites for automotive applications: a review. Polym Adv Technol 33(10):3023–3048
Bodin A, Bäckdahl H, Fink H, Gustafsson L, Risberg B, Gatenholm P (2007) Influence of cultivation conditions on mechanical and morphological properties of bacterial cellulose tubes. Biotechnol Bioeng 97:425–434
Brown AJ (1886) XIX.-The chemical action of pure cultivations of bacterium aceti. J Chem Soc Trans 49:172–187
Bulatović VO, Grgić DK, Slouf M, Ostafinska A, Dybal J, Jozinović A (2019) Biodegradability of blends based on aliphatic polyester and thermoplastic starch. Chem Pap 73(5):1121–1134
Carpenter AW, de Lannoy C-F, Wiesner MR (2015) Cellulose nanomaterials in water treatment technologies. Environ Sci Technol 49(9):5277–5287
Chenab KK, Sohrabi B, Jafari A, Ramakrishna S (2020) Water treatment: functional nanomaterials and applications from adsorption to photodegradation. Mater Today Chem 16:100262
Coskun S (2016) The applications of magnetic nanoparticles in cancer diagnosis [Graduation Thesis, BezmialemVakif University]. Istanbul
Das O et al (2023) Functionalised biochar in biocomposites: the effect of fire retardants, bioplastics and processing methods. Compos Part C: Open Access 11:100368
Dikshit PK, Kumar J, Das AK, Sadhu S, Sharma S, Singh S, Gupta PK, Kim BS (2021) Green synthesis of metallic nanoparticles: applications and limitations. Catalysts 11(8):902
Dobrovolskaia MA, Mcneil SE (2007) Immunological properties of engineered nanomaterials. Nat Nanotechnol 2(8):469–478. https://doi.org/10.1038/nnano.2007.223
Donini ÍAN, Salvi DTB, Fukumoto FK, Lustri WR, Barud HS, Marchetto R (2010) Biosynthesis and recent advances in production of bacterial cellulose. Eclética Quím 35:165–178

Eckel F, Van Opdenbosch D, Schandl KS et al (2024) Dependency of tensile properties and biodegradation on molecular mass during hydrolysis of poly(butylene succinate). npj Mater Degrad 8:97

Falcão SC, Coelho ARDB, Evêncio Neto J (208) Biomechanical evaluation of microbial cellulose (Zoogloea sp.) and expanded polytetrafluoroethylene membranes as implants in repair of produced abdominal wall defects in rats. Acta Cir Bras 23:184–191

Fontana JD, De Souza AM, Fontana CK, Torriani IL, Moreschi JC, Gallotti BJ, Farah LFX (1990) Acetobacter cellulose pellicle as a temporary skin substitute. Appl Biochem Biotechnol 24–25:253–264

Gao XH, Cui YY, Levenson RM, Chung LWK, Nie SM (2004) Cancer targeting and imaging with semiconductor quantum dots. Nat Biotechnol 22(8):969–976

Gao C, Yan T, Du J, He F, Luo H, Wan Y (2014) Introduction of broad spectrum antibacterial properties to bacterial cellulose nanofibers via immobilising ε-polylysine nanocoatings. Food Hydrocoll 36:204–211

García-Guzmán L, Cabrera-Barjas G, Soria-Hernández CG, Castaño J, Guadarrama-Lezama AY, Rodríguez Llamazares S (2022) Progress in starch-based materials for food packaging applications. Polysaccharides 3(1):136–177

Gomez MH, Aguilera JM (1984) A physicochemical model for extrusion of corn starch. J Food Sci 49(1):40–43

Guesh K, Mayoral A, Marquez-Alvarez C, Chebude Y, Diaz I (2016) Enhanced photocatalytic activity of TiO2 supported on zeolites tested in real wastewaters from the textile industry of Ethiopia. Microporous Mesoporous Mater 225:88–97

Kaya GG, Deveci H (2022) Design and synthesis of metal oxide–polymer composites. In: Renewable polymers and polymer-metal oxide composites, pp 101–128. ISBN 9780323851558

Gutiérrez TJ, Ollier R, Alvarez VA (2018) Surface properties of thermoplastic starch materials reinforced with natural fillers. Editorial Springer International Publishing, Functional biopolymers, pp 131–158

Güzel M (2024) Characterization of cellulose produced by bacteria isolated from different vinegars. Int J Biol Macromol 277(3):134436. https://doi.org/10.1016/j.ijbiomac.2024.134436

Hasanin M, Al Abboud MA, Alawlaqi MM, Abdelghany TM, Hashem AH (2022) Ecofriendly synthesis of biosynthesized copper nanoparticles with starch-based nanocomposite: antimicrobial, antioxidant, and anticancer activities. Biol Trace Elem Res 1–14

Hatami H (2013) Importance of water and water-borne diseases: on the occasion of the world water day (March 22, 2013). Int J Prev Med 4(3):243

He CB, Liu DM, Lin WB (2015) Nanomedicine applications of hybrid nanomaterials built from metal-ligand coordination bonds: nanoscale metal-organic frameworks and nanoscale coordination polymers. Chem Rev 115(19):11079–11108

Hellweg S et al (2023) Life-cycle assessment to guide solutions for the triple planetary crisis. Nat Rev Earth Environ 4(7):471–486

Henglein A (1989) Small-particle research—physicochemical properties of extremely small colloidal metal and semiconductor particles. Chem Rev 89(8):1861–1873. https://doi.org/10.1021/cr00098a010

Hijrah N, Syakir MI, Syah BI (2023) Quantification of environmental impact potentials within the processes of industrial wastewater treatment system. IOP Conf Ser: Earth Environ Sci 1167(1):012019

Huang XH, El-Sayed IH, Qian W, El-Sayed MA (2006) Cancer cell imaging and photothermal therapy in the near-infrared region by using gold nanorods. J Am Chem Soc 128(6):2115–2120

Iguchi M, Yamanaka S, Budhiono A (2000) Bacterial cellulose—a masterpiece of nature's arts. J Mater Sci 35:261–270

Imamura K, Yoshikawa T, Hashimoto K, Kominami H (2013) Stoichiometric production of aminobenzenes and ketones by photocatalytic reduction of nitrobenzenes in secondary alcoholic suspension of titanium (IV) oxide under metal-free conditions. Appl Catal B 134:193–197

Jaleh B, Fakhri P (2016) Infrared and fourier transform infrared spectroscopy for nanofillers and their nanocomposites. In: Spectroscopy of polymer nanocomposites. William Andrew Publishing, pp 112–129

Jaouahar M et al (2024) Preparation and characterization of sulfated nanocellulose: from hydrogels to highly transparent films. Int J Biol Macromol 260:129464

Jebali A, Hekmatimoghaddam S, Behzadi A, Rezapor I, Mohammadi BH, Jasemizad T, Sayadi M (2013) Antimicrobial activity of nanocellulose conjugated with allicin and lysozyme. Cellulose 20:2897–2907

Jiang N, Shang R, Heijman SG, Rietveld LC (2018) High-silica zeolites for adsorption of organic micro-pollutants in water treatment: a review. Water Res 144:145–161

Jiang T, Duan Q, Zhu J, Liu H, Yu L (2020) Starch-based biodegradable materials: challenges and opportunities. Adv Ind Eng Polym Res 3(1):8–18

Kamali M, Hewage K, Sadiq R (2019) Conventional versus modular construction methods: a comparative cradle-to-gate LCA for residential buildings. Energy Build 204:109479

Kamari S, Shahbazi A (2021) High–performance nanofiltration membrane blended by Fe3O4@ SiO2–CS bionanocomposite for efficient simultaneous rejection of salts/heavy metals ions/dyes with high permeability, retention increase and fouling decline. Chem Eng J 417:127930

Kausar A, Ahmad I, Maaza M, Eisa MH, Bocchetta P (2022) Polymer/Fullerene nanocomposite for optoelectronics—moving toward green technology. J Compos Sci 6(12):393

Kim SH, Lee CM, Kafle K (2013) Characterization of crystalline cellulose in biomass: basic principles, applications, and limitations of XRD, NMR, IR, Raman, and SFG. Korean J Chem Eng 30:2127–2141

Kim JY, Kim JN, Wee YJ, Park DH, Ryes HW (2007) Bacterial cellulose production by cluconacetobacter sp. RKY5 in a rotary biofilm contractor. Appl Biochem Biotechnol 137–140(1–12):529–537

Klemm D, Schumann D, Udhardt U, Marsch S (2001) Bacterial synthesized cellulose: artificial blood vessels for microsurgery. Prog Polym Sci 26:1561–1603

Klemm D, Heublein B, Fink HP, Bohn A (2005) Cellulose: fascinating biopolymer and sustainable raw material. Angew Chemie—International Ed 44(22):3358–3393

Kumar P, Kumar M (2022) Natural nano-fillers materials for the bio-composites: a review'. J Indian Chem Soc 99(10):100715

Kundu S, Karak N (2022) Polymeric photocatalytic membrane: An emerging solution for environmental remediation. Chem Eng J 438:135575. (42). Zafar R, Zia KM, Tabasum S, Jabeen F, Noreen A, Zuber M (2016) Polysaccharide based bionanocomposites, properties and applications: a review. Int J Biol Macromol 92:1012–1024

Kunioka M, Ninomiya F, Funabashi M (2007) Novel evaluation method of biodegradabilities for oil based poly caprolactone by naturally occurring radiocarbon-14 concentration using accelerator mass spectrometry based on ISO 14855–2 in controlled compost. Polym Degrad Stab 92:1279–1288

Li Y-H, Ding J, Luan Z, Di Z, Zhu Y, Xu C, Wu D, Wei B (2003) Competitive adsorption of Pb2+, Cu2+ and Cd2+ ions from aqueous solutions by multiwalled carbon nanotubes. Carbon 41(14):2787–2792

Li R, Zhang L, Wang P (2015) Rational design of nanomaterials for water treatment. Nanoscale 7(41):17167–17194

Liu WC, Halley PJ, Gilbert RG (2010) Mechanism of degradation of starch, a highly branched polymer, during extrusion. Macromolecules 43(6):2855–2864

Lu C, Su F, Hu S (2008) Surface modification of carbon nanotubes for enhancing BTEX adsorption from aqueous solutions. Appl Surf Sci 254(21):7035–7041

Lu H, Wang J, Stoller M, Wang T, Bao Y, Hao H (2016) An overview of nanomaterials for water and wastewater treatment. Adv Mater Sci Eng 2016

Lv Q et al (2024) Nanocellulose-based nanogenerators for sensor applications: a review. Int J Biol Macromol 259:129268

Madhumitha G, Fowsiya J, Mohana Roopan S, Thakur VK (2018) Recent advances in starch–clay nanocomposites. Int J Polym Anal Charact 23(4):331–345
Mndlovu H, Kumar P, du Toit LC et al (2024) A review of *biomaterial* degradation assessment approaches employed in the biomedical field. npj Mater Degrad 8:66
Mohamed EF, Mohamed F, El-Mekawy A et al (2023) Development of PVA/GO nanocomposites membranes for air-filtration and purification. J Inorg Organomet Polym 33:3389–3401
Mohanty F, Swain SK (2017) Bionanocomposites for food packaging applications. In: Nanotechnology applications in food. Academic Press, pp 363–379
Moon G-H, Kim D-H, Kim H-I, Bokare AD, Choi W (2014) Platinum-like behavior of reduced graphene oxide as a cocatalyst on TiO2 for the efficient photocatalytic oxidation of arsenite. Environ Sci Technol Lett 1(2):185–190
Moradi M, Haghighi M, Allahyari S (2017) Precipitation dispersion of Ag–ZnO nanocatalyst over functionalized multiwall carbon nanotube used in degradation of Acid Orange from wastewater. Process Saf Environ Prot 107:414–427
Mura S, Nicolas J, Couvreur P (2013) Stimuli-responsive nanocarriers for drug delivery. Nat Mater 12(11):991–1003 Moreover, it facilitates the fabrication of synthetic biomaterials for therapeutic purposes and tissue replacement (107)
Muscat D, Adhikari B, Adhikari R, Chaudhary DS (2012) Comparative study of film forming behaviour of low and high amylose starches using glycerol and xylitol as plasticizers. J Food Eng 109(2):189–201
Naushad M, Ahamad T, Al-Sheetan KM (2021) Development of a polymeric nanocomposite as a high performance adsorbent for Pb (II) removal from water medium: equilibrium, kinetic and antimicrobial activity. J Hazard Mater 407:124816
de Oliveira D, Beatrice CAG (2019) Polymer nanocomposites with different types of nanofiller. IntechOpen. https://doi.org/10.5772/intechopen.81329
Otrisal P, Obsel V, Buk J, Svorc L (2018) Preparation of filtration sorptive materials from nanofibers, bicofibers, and textile adsorbents without binders employment. Nanomaterials 8(8):564
Pande VV, Sanklecha VM (2017) Bionanocomposite: a review. Austin J Nanomed Nanotechnol 5(1):1045
Pandey S, Do JY, Kim J, Kang M (2020) Fast and highly efficient catalytic degradation of dyes using κ-carrageenan stabilized silver nanoparticles nanocatalyst. Carbohyd Polym 230:115597
Panthapulakkal S, Sain M (2012) Preparation and characterization of cellulose nanofibril films from wood fibre and their thermoplastic polycarbonate composites. Int J Polym Sci 1–6
Peng X, Li Y, Luan Z, Di Z, Wang H, Tian B, Jia Z (2003) Adsorption of 1, 2-dichlorobenzene from water to carbon nanotubes. Chem Phys Lett 376(1–2):154–158
Perez-Pacheco E, Canto-Pinto JC, Moo-Huchin VM, Estrada-Mota IA, Estrada-León RJ, Chel-Guerrero L (2016) Thermoplastic starch (TPS)-cellulosic fibers composites: mechanical properties and water vapor barrier: a review. In: Composites from renewable and sustainable materials. INTEACH, pp 85–105
Petros RA, DeSimone JM (2010) Strategies in the design of nanoparticles for therapeutic applications. Nat Rev Drug Discovery 9(8):615–627
Phukan M, Nongkhlaw FT (2022) Starch: an overview of its conventional and unconventional sources, pharmaceutical and industrial applications. Curr Trends Pharm Res 9(1)
Picheth GF, Pirich CL, Sierakowski MR, Woehl MA, Sakakibara CN, de Souza CF, de Freitas RA (2017) Bacterial cellulose in biomedical applications: a review. Int J Biol Macromol 104:97–106
Puiggalí J, Katsarava R (2017) Bionanocomposites. In: Clay-polymer nanocomposites. Elsevier, pp 239–272
Qiao K, Zheng YD, Guo SL, Tan J, Chen XH, Li JQ (2015) Hydrophilic nanofiber of bacterial cellulose guided the changes in the micro-structure and mechanical properties of nf-BC/PVA composites hydrogels. Compos Sci Technol 118:47–54
Raghavan D (2006) Charecterization of biodegradable plastics. J Polym-Plast Technol Eng 34–1:41–63

Rahim M, Mas Haris MRH, Saqib NU (2020) An overview of polymeric nano-biocomposites as targeted and controlled-release devices. Biophys Rev 12(5):1223–1231. https://doi.org/10.1007/s12551-020-00750-0
Raja PMV, Barron AR (2024) Rice University. Physical methods in chemistry and nano science. Open Education Resource (OER) LibreTexts Project. https://LibreTexts.org
Portela R, Leal CR, Almeida PL, Sobral RG (2019) Bacterial cellulose: a versatile biopolymer for wound dressing applications. Microb Biotechnol 12(4):586–610. Wiley
Reisz JA, Bansal N, Qian J, Zhao W, Furdui CM (2014) Effects of ionizing radiation on biological molecules—mechanisms of damage and emerging methods of detection. Antioxid Redox Signal 21(2):260–292
Ren X, Chen C, Nagatsu M, Wang X (2011) Carbon nanotubes as adsorbents in environmental pollution management: a review. Chem Eng J 170(2–3):395–410
Rhim JW, Ng PK (2007) Natural biopolymer-based nanocomposite films for packaging applications. Crit Rev Food Sci Nutr 47(4):411–433
Rhim JW, Park HM, Ha CS (2013) Bio-nanocomposites for food packaging applications. Prog Polym Sci 38(10–11):1629–1652
Riggsa BC, Adireddy S, Rehm CH, Puli VS, Elupula R, Chrisey DB (2015) Polymer nanocomposites for energy storage applications. Mater Today: Proc 2:3853–3863. https://doi.org/10.1016/j.matpr.2015.08.004
Rutkowska M, Heimowska A, Krasowska K, Janik H (2002) Biodegradability of poly ethylene starch blends in sea water. Pol J Environ Stud 11:267–274
Rutkowska M, Jastrzebska M, Janik H (1998) Biodegradation of poly caprolactone in sea water. React Funct Polym 38:27–30
Ruz-Cruz MA et al (2022) Thermal and mechanical properties of PLA-based multiscale cellulosic biocomposites. J Market Res 18:485–495
Sadegh H, Shahryari GR, Masjedi A, Mahmoodi Z, Kazemi M (2016) A review on carbon nanotubes adsorbents for the removal of pollutants from aqueous solutions
Saifuddin N, Raziah A, Junizah A (2013) Carbon nanotubes: a review on structure and their interaction with proteins. J Chem 2013
Salama A (2023) Novel cellulose derivative containing aminophenylacetic acid as sustainable adsorbent for removal of cationic and anionic dyes. Int J Biol Macromol 253:126687
Sani A, Dahman Y (2009) Improvements in the production of bacterial synthesized biocellulose nanofibres using different culture methods. J Chem Technol Biotechnol 85(2):151–164
Santhosh C, Velmurugan V, Jacob G, Jeong SK, Grace AN, Bhatnagar A (2016) Role of nanomaterials in water treatment applications: a review. Chem Eng J 306:1116–1137
Sarwar MS, Niazi MBK, Jahan Z, Ahmad T, Hussain A (2018) Preparation and characterization of PVA/nanocellulose/Ag nano composite films for antimicrobial food packaging. J Carbohydarte Polymers 184:453–464
Shalem A, Yehezkeli O, Fishman A (2024) Enzymatic degradation of polylactic acid (PLA). Appl Microbiol Biotechnol 108:413
Shameem MM, Sasikanth SM, Annamalai R, Raman RG (2021) A brief review on polymer nanocomposites and its applications. Mater Today: Proc 45(2):2536–2539
Shankar S, Rhim J-W (2016) Polymer nanocomposites for food packaging applications. In: Functional and physical properties of polymer nanocomposites, first edition, Dasari A, Njuguna J (ed). Wiley
Shen Z et al (2012) An overview of research on agricultural non-point source pollution modelling in China. Sep Purif Technol 84:104–111
Singh TP, Chauhan G, Agrawal RK, Mendiratta SK (2018) Response surface modeling and optimization of tomato puree–casein bio-composite films. Iran Polym J 27:861–879
Spiridon I, Popescu MC, Bodârlău R, Vasile C (2008) Enzymatic degradation of some nanocomposites of poly (vinyl alcohol) with starch. Polym Degrad Stabil 93:1884–1890
Steinbuhel (2001) Bacterial cellulose. Biopolymers. Weinheim: Wiley-VCH

Thanapornsin W et al (2024) Preparation and characterization of biocomposite film made of activated carbon derived from microalgal biomass: an experimental design approach for basic yellow 1 removal. S Afr J Chem Eng 47:178–196
The nano-sized width of the bacterial cellulose is identified by transmission electron microscope [79]
Thomas SP, Stephen R, Bandyopadhyay S (2007) Polymer nanocomposites: Preparation, properties and applications. GAK Gummi Fasern Kunststoffe 2
Tran N, Webster TJ (2010) Magnetic nanoparticles: biomedical applications and challenges. J Mater Chem 20(40):8760–8767
Ul-Islam M, Khan T, Park JK (2012) Nanoreinforced bacterial cellulose–montmorillonite composites for biomedical applications. Carbohydr Polym 89:1189–1197
Umar K, Ibrahim MNM, Ahmad A, Rafatullah M (2019) Synthesis of Mn-doped TiO2 by novel route and photocatalytic mineralization/intermediate studies of organic pollutants. Res Chem Intermed 45(5):2927–2945
Vairavapandian D, Vichchulada P, Lay MD (2008) Preparation and modification of carbon nanotubes: review of recent advances and applications in catalysis and sensing. Anal Chim Acta 626(2):119–129
Venkateshaiah A, Padil VVT, Nagalakshmaiah M, Waclawek S, Černík M, Varma RS (2020) Microscopic techniques for the analysis of micro and nanostructures of biopolymers and their derivatives. Polymers 12:512
Verma J, Petru M, Goel S (2024) Cellulose based materials to accelerate the transition towards sustainability. Ind Crops Prod 210:118078
Vishnupriya C et al (2024) 'Evaluation of heavy metal removal and antibiofilm efficiency of biologically synthesized chitosan- silver Nano-bio composite by a soil actinobacterium Glutamicibacter uratoxydans VRAK 24. Int J Biol Macromol 255:128032
Vitta S, Drillon M, Derory A (2010) Magnetically responsive bacterial cellulose: synthesis and magnetic studies. J Appl Phys 108:53905
Weber M, Lott C, Fabricius KE (2006) Sedimentation stress in ascleractinian coral exposed to terrestrial and marine sediments with contrasting physical, organic and geochemical properties. J Exp Mar Biol Ecol 336:18–32
WHO (2017) Safely managed drinking water: thematic report on drinking water 2017
Wippermann J, Schumann D, Klemm D, Kosmehl H, Salehi-Gelani S, Ahlers T (2009) Preliminary results of small arterial substitute performed with a cylindrical biomaterial composed of bacteria cellulose. Eur J Vasc Endovasc Surg 37:592–596
Wu SC, Lia YK (2008) Application of bacterial cellulose pellets in enzyme immobilization. J MolCatal B Enzym 54:103–108
Wunderlich B (2011) Do biopolymers behave the same as synthetic high polymers? J Therm Anal Calorim 106(1):81–84
Zaidi Z, Siddiqui SI, Fatima B, Chaudhry SA (2019) Synthesis of ZnO nanospheres for water treatment through adsorption and photocatalytic degradation: modelling and process optimization. Mater Res Bull 120:110584
Zain AHM, bin Ab Wahab MK, Ismail H (2019) Effect of calcium carbonate incorporation on the properties of low linear density polyethylene/thermoplastic starch blends. J Eng 15(2):97–108
Zhang Y, Rempel C, McLaren D (2014 b) Thermoplastic starch. In: Innovations in food packaging. Academic Press, pp 391–412
Zhang Y, Rempel C, Liu Q (2014 a) Thermoplastic starch processing and characteristics—a review. CritAl Rev Food Sci Nutr 54(10):1353–1370
Zhou J, Ma Y, Ren L, Tong J, Liu Z, Xie L (2009) Preparation and characterization of surface crosslinked TPS/PVA blend films. Carbohyd Polym 76(4):632–638

Chapter 2
Polymeric Nanocomposites and Their Applications

Vrinda S. Punnakkal and E. I. Anila

2.1 Introduction

A polymer is a macromolecule comprising numerous repeating subunits linked by covalent bonds. The word "polymer" arises from Greek, with *polus* denoting "many" and *meros* denoting "parts". Although humans have used polymers long before, their chemical properties were studied only in the nineteenth century. Hermann Staudinger who won Nobel Prize in 1953 proposed the long chain of covalently bonded atomic structure for the polymer

Polymers are a vast and versatile class of materials which form an integral part of everyday life and advanced technology. Their long and repeating molecular chain structure grants them diverse properties like flexibility, toughness, resilience, chemical tolerance, and durability that depend on chain length, degree of branching, and type of monomers used.

Polymers are of two main types: natural and synthetic. Cellulose, proteins, and DNA are natural polymers. These natural polymers are essential to life and development throughout history. Synthetic polymers have greatly enhanced the scope of material science. Advances in polymer chemistry have made progress in the advancement of plastics, elastomers, fibres, and various composite materials, each with unique applications.

This chapter provides a comprehensive overview of polymer composites, detailing their classification, synthesis methods, and characterization techniques. A wide variety of applications that benefit from these materials, from energy applications

V. S. Punnakkal
Optoelectronic and Nanomaterials' Research Laboratory, Department of Physics, Union Christian College, Aluva, Kerala, India

Sree Sankara College, Kalady, Kerala, India

E. I. Anila (✉)
Christ (Deemed to Be University), Bengaluru, Karnataka, India
e-mail: anilaei@gmail.com

P. M. Visakh (ed.), *Engineering Applications of Polymer-Based Nanocomposites*, Advanced Structured Materials 237, https://doi.org/10.1007/978-3-031-93463-6_2

to biocompatible materials in medical devices, are also discussed. Learning about polymer composites helps readers understand how they are used to create efficient and high-performing solutions in various applications.

2.2 Polymeric Nanocomposites

Composites can be divided into three main groups: metal matrix composites, ceramic matrix composites, and polymer matrix composites. Polymer matrix composites or polymer nanocomposites (PNCs) are combinations of two or more materials. It has polymers as the main matrix and fillers as the dispersion phase (Oliveira and Beatrice 2019; Kaya et al. 2022). Even small additions of nanomaterial fillers to PNCs can improve the mechanical, thermal, and barrier properties without losing processability. For optimal performance of nanocomposites, nanoparticles should be well-dispersed within the polymer matrix. This uniform dispersion is important as it maximizes the interfacial area between components which is essential for property enhancements. The reinforcing effect of the material is influenced by the matrix's properties, the type and nature of the filler used, and its concentration. The particle size, aspect ratio, orientation, and dispersion of fillers within the polymer matrix significantly contribute to the overall mechanical performance of the composite. Polymer nanocomposites can also be classified into different categories based on the type of polymer matrix, method of fabrication, and size and kind of nanofiller (Khan et al. 2023).

- **Classification Based on the Size of Nanofillers**

  Nanofillers can be divided into three groups based on their size.

Zero-Dimensional (0D) Nanofillers of Spherical Shape

All the dimensions of these 0D nanofillers should be in nanoscale, i.e. their dimensions, will not be greater than 100 nm. Examples of these types of nanofillers are nanoparticles and quantum dots. Most of them have got spherical shape and are made of metal, ceramic, crystalline, or amorphous materials.

Single-Dimensional (1D) Nanofillers of Fibrous Structure

Single-dimensional nanofillers have linear structures. Examples of 1D nanofillers include nanotubes, nanowires, nanofibres, and nanorods. They have dimensions in only one direction.

Two-Dimensional (2D) Nanofillers

These nanofillers have dimensions in two directions and their surface area-to-volume ratio is high. Examples are layered silicates, graphene, MXene, nanoflakes, nanobelts, etc.

Three-Dimensional (3D) Nanofillers

In nanofillers with 3D structures, all three spatial dimensions exist. The most common 3D nanostructures include clusters of 1D and/or 2D nanoparticles, 3D nanoporous structures, and 3D hierarchical nanostructured materials. Nanoflowers and nanoprisms are 3D nanostructures.

2.3 Classification Based on the Type of Nanofillers

Nanoclay-Based Polymer Nanocomposites

Nanoclays (e.g. montmorillonite, hectorite, saponite) are layered silicates used as fillers in polymers. Nanoclay fillers are advantageous due to their expansive surface area and large aspect ratio, which improves mechanical, thermal, and barrier properties. They are commonly useful in packaging, automotive parts, and flame-retardant applications (Al-Mutairi et al., 2022).

Carbon-Based Nanocomposites

Carbon Nanotubes: They are either single-walled or multi-walled. Their electrical, mechanical, and thermal properties are excellent. Applications include electronics, sensors, and conductive materials.

Graphene: The planar configuration of carbon atoms in graphene gives it outstanding strength, flexibility, and electrical and thermal conductivity. Graphene nanocomposites are used in electronics, coatings, and flexible devices.

Carbon Black and Fullerenes: These fillers aim to enhance electrical conductivity, thermal stability, and UV resistance. Carbon black is useful in rubber and tyre applications.

Metal and Metal Oxide Nanocomposites

Metal Nanoparticles (e.g. Silver, Gold, Copper): These provide antimicrobial properties, and electrical conductivity, and are widely used in electronics, medical devices, and sensors.

Metal Oxide Nanoparticles (e.g. TiO_2, ZnO, Fe_2O_3, Al_2O_3): Metal oxides can improve mechanical properties, thermal stability, and UV resistance. Common applications are in coatings, paints, and protective films.

Polymer Nanofiber Composites

Polymer Nanofibres (e.g. Electrospun Nanofibres): These polymer nanofibres are used as fillers within another polymer matrix to enhance the mechanical and thermal properties. These are used in tissue engineering, wood dressing, and textiles.

Bio-Based Nanofillers

Cellulose Nanocrystals (CNCs) and Nanofibrillated Cellulose (NFC): We can get these bio-based nanofillers from natural resources. They are renewable and biodegradable and can enhance polymers' mechanical, thermal, and barrier properties. Applications of these composites include sustainable food packaging, drug delivery, etc.

Chitosan Nanoparticles: Chitosan, derived from chitin, is used for its biocompatibility, biodegradability, and antimicrobial properties, often in biomedical and packaging applications.

Hybrid Nanofiller-Based Nanocomposites

Hybrid fillers combine two or more types of nanofillers (e.g. carbon nanotubes and metal oxide particles) to optimize the properties of the nanocomposite. This class aims to integrate the benefits of each filler type, such as conductivity from CNTs and UV resistance from metal oxides.

2.3.1 Classification Based on the Type of Polymer Matrix

Thermoplastic Polymer Composites

These nanocomposites use thermoplastic polymers as the matrix. By heating, they can be repeatedly melted and reshaped. Examples of thermoplastic polymer nanocomposites include polypropylene nanocomposites, polyethylene nanocomposites, and polycarbonate nanocomposites.

Thermosetting Polymer Nanocomposites

In this polymer nanocomposites, thermosetting polymers which undergo irreversible curing, are the matrix. Examples of thermosetting polymer nanocomposites include epoxy resin nanocomposites, phenolic resin nanocomposites, unsaturated polyester nanocomposites, and polyurethane nanocomposites

Elastomer-Based Nanocomposites

The matrix in these nanocomposites is elastomers (rubbery polymers) and offers flexibility and elasticity to the composite. Examples are natural rubber nanocomposites, styrene-butadiene rubber (SBR) nanocomposites, and silicone rubber nanocomposites.

Biodegradable or Biopolymer Nanocomposites

The matrix of this class of nanocomposite includes biodegradable or bio-based polymers like polylactic acid, starch, and cellulose as the matrix.

Conductive Polymer Nanocomposites

Here conductive polymers like polyaniline, polypyrrole, and polyethylene dioxythiophene serve as the matrix and are used for their electrical properties

Speciality Polymer Nanocomposites

Speciality and smart polymer materials like polyether ether ketone, liquid crystal polymers, and polyimide form the matrix in these nanocomposites. They are well-suited for structural uses such as gears, automotive components, electronic device casings, and high-strength composites.

2.4 Classification Based on the Preparation Methods

The preparation of polymer nanocomposites can be grouped into the following:

Ex Situ Synthesis: Nanoparticles are synthesized independently and incorporated into a polymeric solution or molten polymer matrix.

In Situ Synthesis: Nanoparticles are synthesized directly within the polymer matrix by adding precursors to the polymer interstices. These precursors are later converted into nanoparticles.

Polymerization with Pre-synthesized Nanoparticles: The polymerization of monomers occurs in the presence of nanoparticles synthesized beforehand.

Simultaneous Synthesis: The polymerization of the monomer and nanoparticle formation from the precursors occurs concurrently. The performance of polymer matrix composites is influenced by several factors, including:

- The fabrication method employed for the nanocomposite.
- The type, orientation, and characteristics of the added materials.
- The extent of blending of the polymer and filler phases.
- The nature of bonding at the matrix–filler interface.
- Nanoparticles' volume fraction.
- The size, shape, and morphology of the fillers.
- The interphase characteristics generated at the polymer–filler interface.

To enhance the properties, nanoparticles' proper dispersion and distribution are essential within the matrix. Poor dispersion leads to particle agglomeration, forming defects and diminishing the material's overall performance. A consistent dispersion of nanoparticles is critical for property enhancement.

The characteristics of the boundary where the polymer matrix meets the filler are another crucial factor influencing nanocomposite performance. Polymer matrix nanocomposites typically exhibit a phase boundary between the two components, forming an interfacial layer at this boundary. The interfacial region's characteristics, composition, and minute structure differ from those of both the polymer and the filler. A strong bond at the interface greatly enhances the overall properties of the nanocomposite (Thomas and Stephen 2007).

Interphase characteristics are ruled by how the polymer matrix interacts with the surface of nanofiller. Properties of nanocomposite can be effectively tailored by optimizing interfacial bond between polymer and nanofillers. Surface energy ratio of filler and polymer significantly influences interaction at interface. High surface area of nanosized particles impacts interfacial phenomena; however, this subsequently affects overall properties of nanocomposite (Khan et al. 2016).

- **Synthesis of Nanocomposites**

Chemical methods or mechanical methods are used to produce the composites. One of the primary challenges in their fabrication is attaining a uniform and consistent distribution of nanoparticles inside the polymer matrix. Nanofillers often tend to cluster, forming micrometre-sized clusters that hinder proper dispersion, ultimately compromising the performance of the nanocomposite (Sivadas et al. 2021).

The most used techniques for fabricating polymer nanocomposites include:

- Intercalation method.
- In situ polymerization.
- Sol-gel method.
- Direct blending of polymer and nanofillers.

Intercalation Method

Intercalation is a common top-down approach for incorporating nanoplatelet-type nanomaterials into a polymer matrix. It typically requires surface modification of the nanoplatelets to ensure their uniform dispersion as plate-like nanofillers. Intercalated morphology is achieved when polymer chains infiltrate the gallery spaces within the layered structure of the nanoplatelets. Two primary methods are used for this dispersion. One is chemical method which employs *in situ polymerization*, where nanofillers are first mixed in a monomer solution. The polymerization process then proceeds, allowing the nanoplatelets to swell in the monomer. As the polymer forms, it intercalates into the layered structure of the nanoplatelets. In essence, the polymer chains grow between the intercalated sheets, creating a uniform dispersion within the matrix. The second method is mechanical method which involves directly adding nanofillers into the polymer through solution mixing. The polymer and the nanofillers are dissolved in the same solvent. When the two solutions are combined, the polymer chains penetrate the nanoplatelet layers, displacing the solvent and achieving intercalation.

Melt Intercalation Method

Melt intercalation method involves blending nanofillers with a polymer matrix at elevated temperatures, and it is widely adopted in industrial applications. Annealing of the polymer matrix and nanoparticles is done under shear or static conditions, facilitating their integration. This method is agreeable with standard industrial processes like extrusion and injection moulding. This technique facilitates the synthesis of polymers that cannot be produced by other intercalation techniques such as in situ polymerization or solution blending. Melt blending is another similar method in which polymer powders or pellets are melted to form a viscid mixture and nanoparticles are then introduced to this mixture. The dispersion of the nanoparticles is enhanced through high-temperature diffusion coupled with a high shear rate. The material obtained can be fabricated into finished shapes using processes like compression moulding, injection moulding, or fibre production.

In Situ Polymerization

In this approach, the nanofiller should be evenly disbursed within the solution containing the monomer earlier to the polymerization starts to evolve so that the polymer matrix is formed round the nanoparticles. Polymerization can be initiated through methods, like applying heat energy (thermal polymerization) or the usage of a suitable initiator (e.g. Stannic chloride in polymerization of vinyl ethers and epoxides). By this method, polymer-grafted nanoparticles may be created that allow the incorporation of high concentrations of nanofillers. Organic modifiers like sodium dodecyl sulphate (SDS) or Triton X-100 may be brought to improve nanoparticle dispersion and take part in the polymerization. Polymers that are either insoluble or thermally unstable can be used for producing polymer nanocomposites by this technique. This technique can also be performed in the absence of solvents. Overall, this method can produce high-performance materials. Mini-emulsion polymerization is another related technique, that involves dispersing nanoscale monomer droplets within a solution. The advantages of this method include controlled particle morphology, strong interfacial adhesion between nanofillers, and high transparency. This method also facilitates the incorporation of more nanofiller concentrations, improved performance of the final product, processing without using a solvent, and the production of covalently bonded nanoparticle functional groups and polymer chains. Additionally, here we can use both thermoset and thermoplastic polymers. However, a major challenge of this method is the tendency of nanoparticles to agglomerate.

Sol-Gel Method

This method is environment-friendly and used for producing hybrid nanocomposites, generating minimal waste. This process is used to create solid materials from small molecules. During this process, a sol (solution) transforms into a gel-like system composed of liquid and solid phases. The resulting morphologies range from discrete particles to continuous polymer networks. The washing step can be eliminated for the starting materials that do not introduce impurities into the final product, further enhancing the method's ecological efficiency (Rane et al. 2018; Pomogailo 2005).

Advantages:

- Initial materials are at the molecular level, ensuring homogeneous end products.
- Produces materials of relatively high purity.
- Allows for control over the degree of porosity.
- Facilitates the preparation of various sizes and is easier to implement.
- Allows synthesis at low temperatures.
- Provides precise control over the chemical composition, making it ideal for creating multicomponent materials.
- Supports the uniform dispersion of small amounts of dopants in the final product.

Disadvantages:

- Reaction times are relatively long.
- The use of organic solvents may pose health risks.

Direct Merging of Polymer Matrix and Nanofillers

This is a simple and top-down method to produce polymer composites, but the proper distribution of nanofillers is difficult compared to other preparation techniques. This approach relies on breaking down amassed nanofillers during the mixing process and is particularly suited to produce polymer nanocomposites. There are two primary techniques for blending polymers with nanofillers:

(i) Melt Blending

Here, the fillers are incorporated into the polymer matrix above its glass-forming temperature, typically without the use of solvents. Shear stress, generated by viscous drag in the polymer melt, breaks down nanofiller aggregates, ensuring uniform dispersion within the polymer matrix.

(ii) Solvent Blending

This technique involves dispersing nanoparticles in a solvent while dissolving the polymer in a co-solvent. The polymer composites are then recovered by solvent evaporation or coagulation. Compared to melt compounding, this method induces lower shear stresses. Dispersing the nanofillers in the solvent by sonication is the initial step. This is to avoid aggregation of nanofillers before mixing. Once prepared through either approach, polymer nanocomposites are shaped into finished products using conventional methods like injection moulding, calendaring, casting, compression moulding, blow moulding, rotational moulding, extrusion moulding, or thermoforming.

- **Characterization of Nanocomposites**

Characterization is essential for evaluating the properties of polymer nanocomposites including analysing the dispersion of fillers within the polymer matrix, understanding the structure and composition influenced by the processing methods, and assessing how surface modifications of fillers impact their dispersion and the overall properties of the composites (Shimelis Tamene Gobena 2024). Various methods can be used in the analysis of nanocomposites. Commonly used techniques include technique using X-rays such as wide-angle X-ray diffraction (WAXD), small-angled X-ray scattering (SAXS), scanning electron microscopy (SEM), and transmission electron microscopy (TEM). WAXD is very useful to investigate the structure of nanocomposites and investigate the penetration of fused polymers into silicate layers. TEM provides insights into the internal structure, phase distribution, and defects with direct visualization. Laser-based Raman spectroscopy has also been successfully used to analyse carbon-based materials. Fourier-transform infrared spectroscopy (FTIR) can provide information about chemical composition of functional group and bonding in nanocomposites.

2.5 X-Ray Diffraction (XRD)

X-rays are produced within an X-ray tube by accelerating electrons from a heated filament (cathode) towards a metal target (usually tungsten) using a high voltage. Upon impact, the electrons undergo rapid deceleration, primarily through two mechanisms: Bremsstrahlung, where the electrons are slowed down by the strong electric field of the target nuclei, and characteristic X-ray emission, where the impact dislodges inner-shell electrons, resulting in the emission of X-rays as outer-shell electrons fill the vacancies. This energy conversion process generates electromagnetic radiation, including X-rays.

When X-rays pass through a sample, the intensity decreases according to the equation $I = I_0 \exp(-\mu x)$, where I_0 represents the initial intensity, μ is the linear absorption coefficient, and x is the sample thickness. The properties of the material and the energy from the source influence the penetration depth of the X-rays. X-rays interact with the sample causing absorption, scattering, and transmission. The scattered part of the radiation interferes with the creation of diffraction patterns. These patterns depend on the angle of incidence and the direction of the diffraction. The terms "dispersion" and "diffraction" are often used as synonyms in this context. For the amorphous polymers, the diffraction pattern shows broad peaks instead of the usual pointed peaks of crystalline materials. X-ray diffraction pattern (XRD) thus obtained can be used for examining non-crystalline and semi-crystalline polymers. It allows the analysis of many microstructural features, such as lattice parameters, crystal defects, crystallographic orientation, and measurement of crystallinity. Quantitative analysis of these patterns can provide useful information about atomic structure, including interatomic distances, structure, and radial distribution functions. Wide-angle X-ray scattering (WAXS) is used to study structures on a scale of about 1 nm, and small-angle X-ray scattering (SAXS) is used to analyse larger structural features, that range from 1 nm to around 400 nm.

Wide-Angle X-Ray Scattering (WAXS)

In wide-angle X-ray diffraction, the X-rays with wavelength λ strike a material at an angle θ, interacting with crystal planes spaced by a distance d. Diffraction occurs only when the X-rays reflected from successive planes constructively interfere and that happens only when the extra path travelled by the rays equals an integral multiple of the wavelength. Bragg's law describes this phenomenon:

$$n\lambda = 2dsin\theta$$

n represents an integer, λ represents the wavelength of the incident X-rays, d denotes the lattice spacing, and θ is the glancing angle. To satisfy Bragg's law and observe reflections from different crystal planes, θ or λ must vary. Two primary techniques for achieving this are the Laue and powder methods. The Laue method exposes a stationary crystal to polychromatic X-rays (a spectrum of wavelengths). Each set of crystal planes selects the suitable wavelength to satisfy the Bragg condition, creating

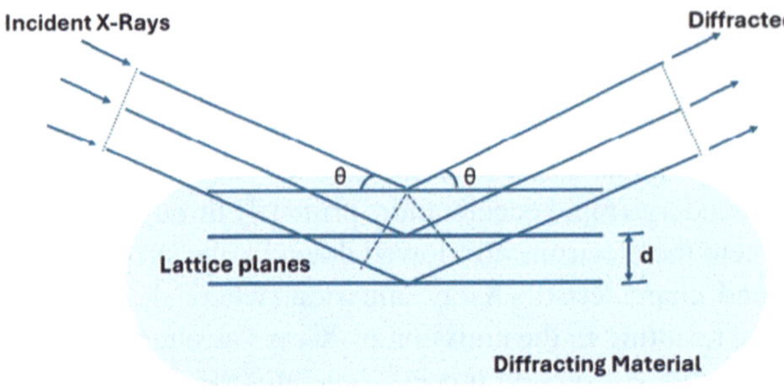

Fig. 2.1 Wide-angle X-ray diffraction

a diffraction pattern that provides insights into the crystal structure. Alternatively, the powder method employs monochromatic X-rays directed at a finely powdered sample. Here, the variability in the Bragg angle θ due to random crystal orientations enables diffraction, as some crystals align appropriately to meet the Bragg condition. The resulting diffraction pattern can reveal structural details of the material (Fig. 2.1).

Small-Angle X-Ray Scattering (SAXS)

Small-angle X-ray diffraction (SAXS) is a valuable method for analysing structural features ranging in size from 1 nm to approximately 400 nm. In typical experiments, X-rays are generated using the CuKα emission line of wavelength $\lambda = 0.154$ nm. The corresponding glancing angles for lattice spacings of $d = 1$ nm and $d = 100$ nm are $\theta = 4.4°$ and $\theta = 0.04^{\circ}$, respectively. The experimental setup for SAXS resembles that used in wide-angle X-ray scattering (WAXS). However, due to the proximity of the scattered beam to the primary transmitted beam, a highly precise collimation of the incident beam is essential. Scattered X-ray intensity is measured as a function of the scattering angle, and the diffraction patterns are obtained from this intensity. By analysing these patterns, one can investigate specific crystal defects by detecting variations in scattered intensity in the regions surrounding the defect (Bergstrom 2015).

2.6 Electron Microscopy (EM)

The significance of electron microscopy (EM) in the field of nanocomposites stems from two main factors. First, it offers highly detailed information, surpassing that provided by other techniques. EM enables the examination of the morphology of the polymer matrix, the filler, and the adhesion between them with nano-metre-level resolution. Second, it facilitates the analysis of how structural components of the composite respond to applied loads, even during in situ experiments, aiding in the development of customized materials. Moreover, electron microscopy uniquely provides clear evidence of the intercalation and exfoliation of fillers within the polymer matrix, allowing precise quantification of the morphological characteristics of polymer nanocomposites (Adhikari and Michler 2009). Electron microscopy is of two types, scanning electron microscopy (SEM) and transmission electron microscopy (TEM).

Scanning Electron Microscopy (SEM)

The scanning electron microscope (SEM) is a powerful imaging tool that uses a focused beam of electrons to produce highly magnified images of samples (Raja and Barron 2024). This analysis does not offer specific details regarding the filler's dispersion, distribution, or interaction with the polymer matrix. However, it does provide insights into whether the filler is uniformly dispersed throughout a significant volume of the material. Additionally, it reveals information about the filler's affinity for the polymer matrix (Adhikari and Michler 2009). Secondary electrons are emitted from the specimen when they get energy from the primary electrons from the source. This is the basic principle on which SEM operates. An image is generated by using the secondary electrons from various points on the sample. SEM requires a vacuum environment for high-resolution imaging to prevent interactions between electrons and gas molecules. Additionally, the primary electrons emitted from the electron gun are accelerated by applying heat or high energy, typically in the 1–40 keV range (Akhtar et al. 2018). For bulk materials' SEM analysis, a common sample type is a cryo-fractured surface. On the other hand, a flat surface can be obtained using microtomy, and in some cases, these microtomed surfaces are chemically etched. Thin films on a substrate can be analysed directly. To prevent surface charging during SEM analysis, a thin layer of carbon is coated on the surface of the polymer (Fig. 2.2).

The enhancement of a polymer nanocomposite's material properties is largely affected by the distribution of fillers within the polymer. During the fabrication process, material scientists focus on improving the bonding between the fillers and the polymer, commonly referred to as the polymer–filler interaction. However, the bonding between fillers and the polymer matrix may not be effective always. Instead, they often exhibit a readiness to bond among themselves, leading to the formation of aggregates, known as filler–filler interactions. These interactions can weaken the properties of the composite material. In SEM grayscale images, these filler clusters or agglomerates are typically observed as white regions. Material scientists frequently assess the characteristics of these clusters manually, but such observations are prone to errors. Earlier research utilizing SEM analysis for polymer blend characterization has primarily focused on directly detecting objects, such as phase domains, within the images. Parameters like the size and shape of these domains have been evaluated using various image-processing software tools (Kunda et al. 2015). A scanning electron microscope (SEM) is often paired with an X-ray detector, such as an energy-dispersive (EDX) or wavelength-dispersive (WDX) system. This allows the characteristic X-ray emissions of specific elements to be analysed. The elemental composition and distribution across a surface can then be visualized on the imaging screen (Michler and Lebek 2016).

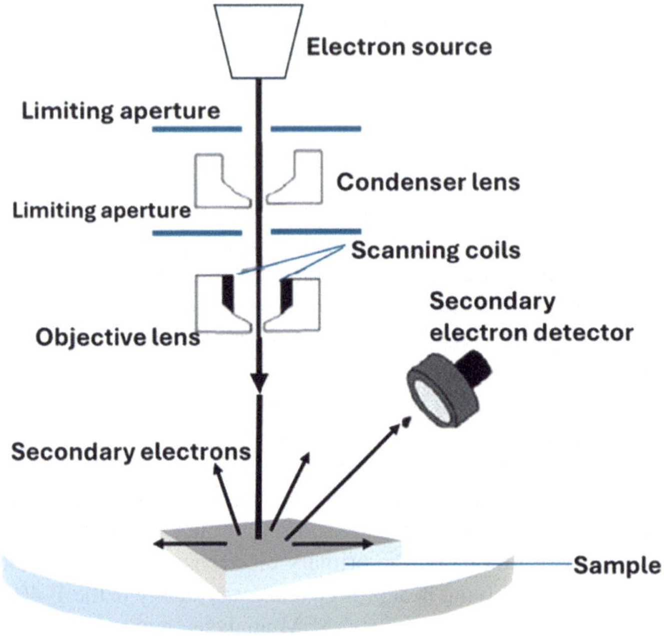

Fig. 2.2 Diagram of scanning electron microscope

Transmission Electron Microscopy (TEM)

TEM is an advanced analytical technique for examining and visualizing materials at the nanoscale. TEM produces much higher-resolution images, capable of revealing minute details, including individual atoms. The term "transmission" reflects the TEM's method of operation, where a beam of high-energy electrons passes through an ultra-thin sample. Certain regions of the sample may block or deflect the electron beam, while others allow it to pass through. Electrons that transmit through the sample are captured beneath it using a camera or a glowing screen and form the image. The darker regions of the image represent the areas where few or no electrons pass through, while brighter regions indicate higher electron transmission. Intermediate grey regions in the image are formed because of the specific interaction of electrons with the sample (Fig. 2.3).

One major challenge of TEM analysis is the preparation of extremely thin samples to allow sufficient electron transmission with minimal energy loss. TEM gives information about particle size, shape, distribution, crystallinity, crystallite size, structure, and orientation (Venkateshaiah et al. 2020). Nanoclays dispersed in various polymer matrices are extensively studied using TEM images and are commonly employed to examine the dispersed structure of the clay. High magnification is needed to assess the distribution of nanoscale fillers, making TEM images a popular choice for this analysis. However, the images typically represent only a limited area, which may not accurately reflect the filler distribution throughout the entire sample, leaving the

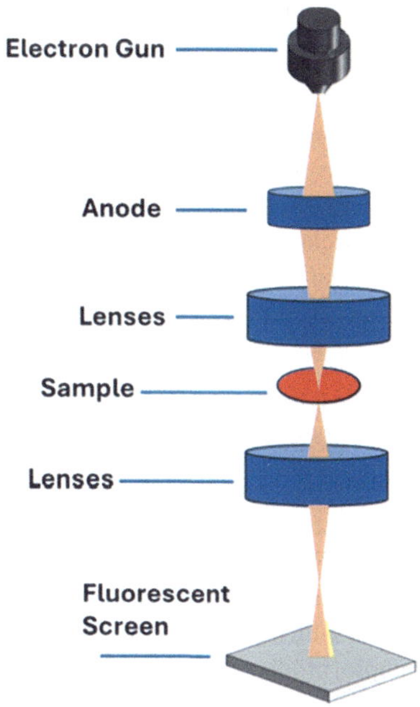

Fig. 2.3 Schematic representation of transmission electron microscope

overall uniformity of the filler distribution uncertain (Kim et al. 2007; Asadabad et al. 2015).

2.7 Atomic Force Microscopy (AFM)

The AFM, also known as Scanning Probe Microscopy, is an important technique to analyse surface topography with a resolution near the atomic scale. It can measure the surface roughness of samples with angstrom-level precision. In AFM, the surfaces of the samples are sensed with a nanoscale sharp tip connected to a flexible cantilever that behaves like a spring. When the tip interacts with the specimen surface, the cantilever bends, and this bending is measured using a laser diode and a split photodetector. The line-by-line scanning of the sample surface is done by this sharp tip, through the specific scanning method which again depends on the selected operating mode. AFM techniques are broadly grouped into two main operating modes: contact mode and dynamic (or tapping) mode. In contact mode, the tip maintains constant contact with the surface, and an electronic feedback system adjusts the tip's position to ensure a consistent interaction force during the raster scan. In dynamic or tapping mode, the cantilever vibrates at resonance frequency and the tip oscillates up and down. The tip comes in contact with the surface of the sample intermittently, and the feedback loop maintains the amplitude of the tapping motion constant. AFM nanomechanics techniques directly characterize key mechanical properties, such as elastic modulus and adhesion energy, on the material's surface. This microscopic information helps the researchers to understand the underlying mechanisms involved in materials. The scale of data obtained through AFM nanomechanics closely aligns with the resolution of computer simulations, offering a bridge between simulation results and the macroscopic properties of materials. These techniques have become invaluable for understanding the mechanical behaviour of polymeric materials. Applying AFM nanomechanics to materials under deformation provides insights into their microscopic deformation behaviour. It also allows for direct visualization of the nanoscale surface stress profile of the sample in real space by observing changes in the modulus.

2.8 Raman Spectroscopy

The chemical composition of a sample can be studied using Raman spectroscopy through Raman scattering. The sample is illuminated with a single wavelength laser light source, and the resulting Raman shifts are analysed. These shifts represent changes in frequency that occur when the incident laser light interacts with the sample, producing scattered light with a frequency different from that of the source. This frequency shift corresponds to the difference in energy between the excited and ground vibrational states of the sample and creates a distinct "fingerprint" for sample identification. Broadband sources cannot be used in Raman spectroscopy since they mask the subtle wavelength differences between the excitation and scattered photons. The wavelength of the light source holds significant importance in the spectroscopy setup since it influences the resolution, signal intensity, and overall cost. Ultraviolet to infrared range lasers used in Raman spectroscopy with each wavelength offering specific advantages and limitations depending on the application (Toncheff and Bishop n.d.). Raman spectroscopy is utilized for analysing organic composites, and it gained renewed interest recently, particularly with the development of carbon nanotubes. This is a non-destructive method and is widely utilized to characterize the vibrational states of different organic materials like diamond, graphite, graphene, etc. Due to these materials' resonance-enhanced Raman scattering effects, they produce strong and distinct bands, even when dispersed in polymer matrices at low concentrations. This makes Raman spectroscopy a crucial tool for studying composites containing carbon materials. When carbon materials are analysed using a 633 nm laser, the Raman spectrum typically shows a G band at around 1580 cm^{-1}, a D band at approximately 1333 cm^{-1} and a high-frequency shoulder near 1600 cm^{-1} (D′ band) characteristic of graphite crystals. These D and D′ bands are associated with structural defects in the carbon, such as edges or distorted layers. The level of disorder is assessed by calculating the I_D/I_G ratio, which represents the intensity ratio of the defect-induced D and G bands. Additionally, the spectrum of carbon materials often includes the G′ band around 2681 cm^{-1}, representing the D band's first overtone. This G′ band may also appear in defect-free carbon structures (Bokobza 2018). For analysing polymer nanocomposites, other often employed techniques include

Fourier-transform infrared (FTIR) spectroscopy, differential scanning calorimetry (DSC), and thermogravimetric analysis (TGA).

2.9 Fourier-Transform Infrared (FTIR) Spectroscopy

The absorption of infrared radiation at various wavelengths can detect functional groups in organic and inorganic compounds. The process begins with the collection of an interferogram from the specimen by an interferometer. To generate the spectrum, the Fourier transform, a mathematical algorithm is applied to the interferogram. Modern FTIR spectrometers can digitize interferograms, perform the Fourier transform, and display the resulting spectrum. They can operate in multiple modes, such as absorption, total and diffuse reflectance, attenuated total reflectance (ATR), and photoacoustic modes, making them suitable for analysing solid, liquid, or gaseous samples (Berna 2017). There are mainly three regions in infrared (IR) spectrum: the far-IR (below 400 cm^{-1}), the mid-IR (4000–400 cm^{-1}), and the near-IR (14,000–4000 cm^{-1}). The mid-IR range is extensively used since it covers the characteristic absorption and emission bands for most compounds. This range is again divided into two: the functional group region (4000–1400 cm^{-1}) and the fingerprint region (1400–400 cm^{-1}). Molecules with similar functional groups may have comparable spectra in the functional group region. However, the fingerprint region typically displays unique spectral patterns that aid in distinguishing different compounds. The vibrations of atoms within a molecule can be categorized into two primary types: stretching and bending. Stretching involves changes in bond length, which can occur symmetrically or asymmetrically while bending refers to changes in bond angles. The sample absorbs the IR radiation if the radiation's frequency matches the vibrational frequency of a particular bond within the molecule. This absorption corresponds to the energy associated with functional groups, making IR spectroscopy a valuable tool for understanding chemical structures. The IR spectrum displays a plot of %transmittance (or absorbance) against the wave number and peaks indicate absorption at particular wave numbers, giving information about the molecule's structure and functional groups. When immiscibility exists between polymers and fillers in a composite, the observed spectrum typically reflects the combined spectra of the individual materials. However, chemical interactions between the polymer chains may occur for miscible components, resulting in changes to the spectra of the nanocomposite compared to those of the pure components. Shifts in wave numbers and the broadening of absorption bands in the spectrum are indications of these interactions. (Jaleh and Fakhri 2016).

2.10 Differential Scanning Calorimetry (DSC)

Differential scanning calorimetry (DSC) is a thermo-analytical method employed to measure the difference in heat flow between a sample and a reference, kept at identical temperatures, as their temperatures are increased during the experiment. In a physical transformation such as a phase change, the heat flow to the sample differs from that of a reference sample maintained at the same temperature. The direction of heat transfer determines whether the process is endothermic or exothermic. When a solid changes into a liquid, it requires additional heat to keep the temperature consistent with the reference due to the latent energy absorbed during the endothermic melting process. During an exothermic process like crystallization, the sample releases heat, so that less amount of heat is needed to maintain its temperature. The sample is characterized by monitoring the variations in the heat requirement and quantifying the heat transfer involved in these transformations. It is a widely utilized technique in industry for quality control and monitoring polymer curing. It effectively probes processes like melting, crystallization, and glass transitions. The method's rapid measurement capabilities and high accuracy make it invaluable for polymer characterization. The studies have shown that while low nanoparticle loadings can effectively nucleate crystallization, higher loadings can disrupt the growth of crystals, hindering the overall crystallization process (Pielichowski and Pielichowska 2018).

2.11 Thermogravimetric Analysis (TGA)

TGA is a method used to monitor variations in a sample's mass relative to time or temperature. The sample is contained in an aluminium container and heated steadily in a controlled environment. Physical and chemical characteristics of the materials are studied by analysing and characterizing different material systems and comparing thermograms of different compounds. Such a method is highly accurate since it can detect mass changes of a fraction of microgram level by weighing fluctuations caused by thermal decomposition. TGA performed on different polymer nanocomposites (PNs) has shown that the incorporation of carbon nanotubes enhances the thermal stability of many polymers, marked by an elevated onset temperature for degradation. In polymer–clay nanocomposites, the enhanced thermal stability is widely attributed to char formation. TGA can provide indirect insights into the extent of nano clay exfoliation within the composite. Recent studies suggest that incorporation of layered silicates into a polymer matrix enhances thermal stability (Corcione et al. 2012).

2.12 Applications of Nanocomposites

There are multiple applications of polymer nanocomposites. It is impossible to cover all of them. Some of the potential applications are described in this section.

2.12.1 Energy Applications

a. Solar Cells

The rising energy demand and the depletion of non-renewable resources like fossil fuels have highlighted the importance of renewable energy. Polymer nanocomposites are an important class of materials with diverse applications due to their excellent properties. The performance of solar cells can be significantly enhanced through the integration of polymers embedded with nanometals, which enhance solar energy absorption. Plasmonic metal nanoparticles exhibit a quantum yield greater than one, making them an excellent choice for solar cell applications. Metal-polymer composites function as metal-dielectric combinations, where collective oscillations of electron clouds, known as Surface Plasmon Resonance (SPR), occur at the interface between metal nanoparticles and the dielectric matrix. These SPR phenomena arise from the interaction between incident light and metal nanoparticles, contributing to the improved performance of solar cells (Shameem et al. 2021).

b. Energy Storage

Ceramics and polymers are being considered for energy storage, but their limitations in dielectric properties, including the dielectric constant in ceramics and the breakdown field in polymers, hinder their application in high-energy-density systems. To address this, composites of polymers and ceramics have garnered significant interest. These composites offer the potential to enhance the dielectric constant and breakdown field simultaneously, resulting in substantial improvements in energy storage capacity (Riggsa et al. 2015). The graphene-polyethylene-dioxythiophene (G-PEDOT) structure significantly enhances the electrochemical performance of PEDOT-based supercapacitors. G-PEDOT facilitates faster electrochemical reactions, achieving an average capacitance of 350 F/g (Alvia et al. 2011). Polysulfone membranes exhibit excellent resistance to acid corrosion and maintain stable mechanical properties even at temperatures exceeding 60°C, making them well-suited for use in hydrogen fuel cells (Ashwin Sidharth, et al. 2022).

2.12.2 Environmental Applications

a. Water Purification

Polymer nanocomposites (PNCs) possess notable properties such as high adsorption capacity, along with catalytic and magnetic functionalities, which make them effective for dye removal. The incorporation of metals creates functional sites, enabling the design of various hybrid structures for this application. For example, metal-decorated polypyrrole nanocomposite anodes have been investigated for the electrochemical oxidation of Direct Red 80, achieving degradation rates ranging from 83% to 100%. Graphene oxide-based polymer nanocomposites are effective adsorbents for removing organic dyes from aqueous solutions. Microorganisms, pesticides, pathogens, and various organic substances are also significant water pollutants. Metal-bound copolymer beads have proven highly effective, achieving up to 99.9% removal of a wide range of bacterial strains. It is reported that AgNP/CNT-coated polyaniline has an excellent antimicrobial and antifouling property.

b. Air Filtration

Polyvinyl alcohol and graphene oxide membranes, incorporating new carbon materials with high surface areas and different oxygen-based functional groups, have shown great capacity for adsorbing pollutants such as formaldehyde and sulphur dioxide gas. The adsorption efficiency of these pollutants is influenced by the quantity, surface area, and functional groups present in the GO membrane. Laboratory tests revealed a 90% removal rate for formaldehyde, while indoor experiments achieved a 75% removal rate. This PVA/GO nanocomposite filter is a promising solution for treating indoor air contaminated with gaseous pollutants and holds potential for various industrial applications (Mohamed et al. 2023).

2.12.3 Biomedical Applications

The usage of biopolymers for extended release of drugs is a widely recognized approach. Nano-bio composites can act as devices for sustained or targeted drug delivery due to their advantages, including biodegradability, biocompatibility, safety, environmental friendliness, and affordability. Recently several nano-bio composite-based drug delivery systems have been introduced commercially. Among biopolymers, starch-based materials such as alginate and chitosan are currently considered ideal for the synthesis of nano-bio composites. In situ solution polymerization and emulsion polymerization are used for producing these nano-biocomposites (Rahim et al. 2020). Merging bioactive compounds with nanotechnology offers a highly innovative strategy for cancer treatment. Natural bioactive substances, in addition to their nutritional benefits, can serve as chemotherapeutic agents for managing cancer. Biopolymers such as alginate, starch, xanthan gum, pectin, guar gum, hyaluronic acid, gelatine, albumin, chitosan, cellulose, and collagen have been effectively utilized to deliver remedial agents to targeted sites. Natural polymer-based nano-biocomposites are used to develop advanced cancer-fighting drug administration systems because of their biodegradability. While there are various methods for creating biopolymer-based nanoparticle systems, it is essential to develop these systems in a cost-effective and environmentally sustainable manner. This ensures they are more accessible, selective, and less toxic, ultimately enhancing treatment efficacy (Mondal et al. 2023).

2.12.4 Electronics and Optoelectronics

Intrinsically conducting polymers like polyaniline, polypyrrole, polythiophene, and poly (3,4-ethylenedioxythiophene) are commonly utilized in optoelectronic devices. To improve their properties, nanocarbon particles are often integrated into conjugated polymers. The performance of polymer/fullerene composites in optoelectronics is heavily influenced by the distribution of fullerene and its interactions with the polymer matrix. Additionally, non-conjugated polymers have also found applications in optoelectronic systems. The advancement of polymer/fullerene nanocomposites in optoelectronics relies on innovative designs, enhanced fullerene dispersion, and its functional integration within conjugated polymers. Effective fabrication techniques, optimized polymers, and functional fullerene nanoparticles are essential for improving the performance of all types of optoelectronic devices. Although green polymer/fullerene nanocomposites currently demonstrate good performance, further developments are necessary for the creation of sustainable green LEDs, OLEDs. Recently significant progress has been made in material design and fabrication processes, leading to improved integration of optoelectronic nanomaterials and enhanced device architectures. Fullerene-based green nanocomposites are considered promising for the next generation of optoelectronics, enabling the development of flexible and miniaturized devices. Recent advancements emphasize the potential of thiophene-based green polymers and eco-friendly synthesized or functionalized fullerenes for applications in energy conversion, sensing, and diode devices. Utilizing green nanocomposites processed through environmentally friendly solution-based methods offers an effective route to achieve optimal results in this field (Kausar et al. 2022).

Food Packaging

Polymer nanocomposites exhibit various properties including transparency, lightweight structure, UV-light blocking, and thermal and dimensional stability that are beneficial for packaging applications. When nanofillers within the polymer matrix are smaller than the wavelength of visible light, then they do not obstruct light transmission, allowing the films to remain transparent. Additionally, when these clay platelets are uniformly intercalated or exfoliated within the matrix, they offer excellent UV-light barrier properties. Studies have shown that agar-based bio-nanocomposite films with copper nanoparticles exhibit strong UV-light blocking capabilities without compromising the clarity of the agar film. These nanocomposites with transparency and UV-light barrier properties are well-suited for packaging food or coating applications that require optical clarity and UV protection. These materials are suitable for packaging processed meats, cheeses, confectionery, cereals, fruit juices, and dairy products, used in beer and carbonated drink bottles and containers, and coatings for paper packaging.

Thermal properties of the biopolymer films are also to be considered for large-scale production of thermoformable films that must retain dimensional stability during high-temperature processing. The addition of nanofillers such as layered silicate clays, cellulose nanofibres, metallic nanoparticles, and chitin nanofibrils into biopolymer greatly enhances thermal and dimensional stability, and this is due to the nanofillers' higher modulus and lower thermal expansion coefficient than the polymer matrix. Thus, the films will get greater stability against shrinkage and deformation during the processing of food packaging materials (Shankar and Rhim 2016).

Sensors

Sensors play an important role in detecting environmental and biological changes, with applications ranging from environmental monitoring to medical diagnostics. Polymer nanocomposites are effective as sensors because of their ability to excite surface plasmon resonance, a key property of metal nanoparticles. The polymer composites are used in gas and vapour sensing applications since they can detect a wide variety of hazardous gases and vapours. They are also used in agriculture applications as humidity sensors. By combining the properties of metals with polymers, these composites offer interdisciplinary applications making these materials suitable for biosensors, including devices for detecting viruses and cancer molecules (Muhammed Shameem et al. 2021). A major challenge of the polymer-based sensors is in achieving consistent and reproducible results.

2.13 Conclusion

Polymer nanocomposites are recognized as a breakthrough category of materials because of their distinctive properties and the very wide range of applications. Different possibilities of synthesis methodologies, characterization approaches, and a broad range of applications of polymer nanocomposites have been discussed in this chapter.

The incorporation of nanofillers within polymer matrices makes them exhibit enhanced physical, chemical, thermal, mechanical, and optical properties and find applications in areas such as the packaging, electronics, energy storage, water purification, and biomedical engineering sectors. Improvements in synthesis methodologies—in situ polymerization, solution blending, and melt mixing—permit the very tight control of distribution and interaction at the nano level within a polymer matrix, resulting in improved properties. Characterization methodologies, such as electron microscopy, X-ray diffraction, thermal analysis, and various spectroscopic techniques, can cover relations between structure and properties that define the material behaviour.

Polymer nanocomposites have become important with the increase in demand for environmentally friendly, high-performance materials with reduced weight and well-balanced eco-sustainability. New eco-friendly synthesis pathways, biopolymer-based nanocomposite development, and functionalized nanofillers bring the industry closer to taking care of environmental issues while propelling advancement in the field of material science. Several issues, such as scalability, cost reduction, and uniform dispersion of fillers, are yet to be improved.

The polymer nanocomposites offer a bridge between pure polymers and advanced materials. As such, they offer an opening of new frontiers for scientific exploration and industrial applications. Their applicability and flexibility make them indispensable for future technologies, driving innovation in different fields and towards a sustainable and innovative future.

References

Adhikari R, MichlerGH (2009) Polymer nanocomposites characterization by microscopy. J Macromol Sci 49:141–180. https://doi.org/10.1080/15583720903048094

Akhtar K, Khan SA, Khan SB, Asiri AM (2018) Scanning electron microscopy: principle and applications in nanomaterials characterization. In: Sharma S (eds) Handbook of materials characterization. Springer, Cham. https://doi.org/10.1007/978-3-319-92955-2_4

Al-Mutairi NH, Mehdi AH, Kadhim BJ (2022) Nanocomposites materials definitions, types and some of their applications: a review. Eur J Res Dev Sustain 3:102–108. https://www.scholarzest.com

Alvia F, Ram MK, Basnayaka PA, Stefanakos E, Goswami Y, Kumar A (2011) Graphene–polyethylenedioxythiophene conducting polymer nanocomposite based supercapacitor. Electrochimica Acta 56:9406–9412. https://doi.org/10.1016/j.electacta.2011.08.024

Asadi Asadabad M, Eskandari M (2015) Transmission electron microscopy as best technique for characterization in nanotechnology. Synth React Inorg 45. https://doi.org/10.1080/15533174.2013.831901

Ashwin Sidharth N et al (2022) IOP Conf Ser: Mater Sci Eng 1219:012042

Bergstrom J (2015) Experimental characterization techniques, mechanics of solid. Polymers 19–114. https://doi.org/10.1016/B978-0-323-31150-2.00002-9

Berna F (2017) Fourier transform infrared spectroscopy (FTIR). In: Gilbert AS (eds) Encyclopedia of geoarchaeology. Encyclopedia of earth sciences series. Springer, Dordrecht. https://doi.org/10.1007/978-1-4020-4409-0_15

Bokobza L (2018) Spectroscopic techniques for the characterization of polymer nanocomposites: a review. Polymers 10. https://doi.org/10.3390/polym10010007

Corcione CE, Frigione M (2012) Characterization of nanocomposites by thermal analysis. Materials 5:2960–2980. https://doi.org/10.3390/ma5122960

Jaleh B, Fakhri P (2016) Infrared and Fourier transform infrared spectroscopy for nanofillers and their nanocomposites. In: Spectroscopy of polymer nanocomposites. William Andrew Publishing, pp 112–129

Kausar A, Ahmad I, Maaza M, Eisa MH, Bocchetta P (2022) Polymer/fullerene nanocomposite for optoelectronics—moving toward green technology. J Compos Sci 6(12):393. https://doi.org/10.3390/jcs6120393

Kaya GG, Deveci H (2022) Design and synthesis of metal oxide–polymer composites. In: Renewable polymers and polymer-metal oxide composites, pp 101–128. ISBN 9780323851558. https://doi.org/10.1016/B978-0-323-85155-8.00005-4
Khan WS, Hamadneh NN, Khan WA (2016) Polymer nanocomposites–synthesis techniques, classification and properties. Sci Appl Tailored Nanostructures 50:50–66
Khan A, Iqbal S, Khan M, Sarwar A (2023) A comprehensive review on polymer nanocomposites; classification, properties and potential applications. J Nanoscope 4:45–74. https://doi.org/10.52700/jn.v4i1.88
Kim D, Lee JS, Barry CM, Mead JL (2007) Microscopic measurement of the degree of mixing for nanoparticles in polymer nanocomposites by TEM images. Microsc Res Tech 70(6):539–546. https://doi.org/10.1002/jemt.20478. PMID: 17479987
Kundu S, Jana P, De D, Roy M (2015) SEM image processing of polymer nanocomposites to estimate filler content. In: IEEE international conference paper. https://doi.org/10.1109/ICECCT.2015.7226104
Michler GH, Lebek W (2016) Chapter 3 electron microscopy of polymers. In: Polymer morphology: principles, characterisation, and processing. https://doi.org/10.1002/9781118892756.ch3
Mohamed EF, Mohamed F, El-Mekawy A et al (2023) Development of PVA/GO nanocomposites membranes for air-filtration and purification. J Inorg Organomet Polym 33:3389–3401. https://doi.org/10.1007/s10904-023-02762-1
Mondal A, Nayak AK, Chakraborty P, Banerjee S, Nandy BC (2023) Natural polymeric nanobiocomposites for anti-cancer drug delivery therapeutics: a recent update. Pharmaceutics 15(8):2064. https://doi.org/10.3390/pharmaceutics15082064
de Oliveira D, Beatrice AG (2019) Polymer nanocomposites with different types of nanofiller. IntechOpen. https://doi.org/10.5772/intechopen.81329
Pielichowski K, Pielichowska K (2018) Polymer nanocomposites. In: Vyazovkin S, Koga N, Schick C (eds) Handbook of thermal analysis and calorimetry. Elsevier Science B.V., pp 431–485
Pomogailo AD (2005) Polymer sol–gel synthesis of hybrid nanocomposites. Colloid J 67:658–677. https://doi.org/10.1007/s10595-005-0148-7
Rahim M, Mas Haris MRH, Saqib NU (2020) An overview of polymeric nano-biocomposites as targeted and controlled-release devices. Biophys Rev 12(5):1223–1231. https://doi.org/10.1007/s12551-020-00750-0
Raja PMV, Barron AR (2024) Physical methods in chemistry and nano science. Open education resource (OER) LibreTexts Project. Rice University https://LibreTexts.org
Rane AV, Kanny K, Abitha VK, Thomas S (2018) Methods for synthesis of nanoparticles and fabrication of nanocomposites. In: Synthesis of inorganic nanomaterials. https://doi.org/10.1016/B978-0-08-101975-7.00005-1
Riggsa BC, Adireddy S, Rehm CH, Puli VS, Elupula R, Chrisey DB (2015) Polymer nanocomposites for energy storage applications. Mater Today: Proc 2:3853–3863. https://doi.org/10.1016/j.matpr.2015.08.004
Shameem MM, Sasikanth SM, Annamalai R, Raman RG (2021) A brief review on polymer nanocomposites and its applications. Mater Today: Proc, Volume 45(2):2536–2539. https://doi.org/10.1016/j.matpr.2020.11.254
Shankar S, Rhim J-W (2016) Polymer nanocomposites for food packaging applications. Functional and physical properties of polymer nanocomposites, first edition. Dasari A, Njuguna J (ed). Wiley. https://doi.org/10.1002/9781118542316.ch3
Gobena ST, Woldeyonnes AD (2024) A review of synthesis methods, and characterization techniques of polymer nanocomposites for diverse applications. Discov Mater 4:52. https://doi.org/10.1007/s43939-024-00119
Sivadas BO, Ashcroft I, Khlobystov AN, Goodridge RD (2021) Laser sintering of polymer nanocomposites. Adv Ind Eng Polym Res 4:277–300. https://doi.org/10.1016/j.aiepr.2021.07.003

Thomas SP, Stephen R, Bandyopadhyay S (2007) Polymer nanocomposites: preparation, properties and applications. GAK Gummi Fasern Kunststoffe 2:49–56. https://www.researchgate.net/publication/233950668

Toncheff C, Bishop E (n.d.) Basic principles of Raman scattering and spectroscopy

Venkateshaiah A, Padil VVT, Nagalakshmaiah M, Waclawek S, Černík M, Varma RS (2020) Microscopic techniques for the analysis of micro and nanostructures of biopolymers and their derivatives. Polymers 12:512. https://doi.org/10.3390/polym12030512

Chapter 3
Starch-Based Nanocomposite for Food Packaging Applications

Aquif Suleman, Jyotismita Bora, Amar Jyoti Kalita, and Bolin Chetia

Abstract Starch-based nanocomposites have become viable alternatives to petroleum-based plastics in food packaging due to the surge in demand for sustainable and environment friendly materials. A comprehensive overview of starch is provided in this chapter, emphasizing its natural abundance, renewability, and film-forming properties. The insufficiency of mechanical strength as well as high water sensitivity of pure starch films render them unsuitable for food packaging applications. Starch-based nanocomposites have been identified as promising materials to address these challenges by incorporating nanofillers like cellulose nanofibers, clay nanoparticles, metal oxides, and biopolymer blends. A variety of structural modifications and nanomaterial reinforcements are explored in this chapter. They also possess antimicrobial, oxygen, and moisture barriers, which are critical to lengthening the viability duration of packaged foods. Furthermore, starch-based composites also ensure food quality and safety while reducing plastic waste.

Keywords Starch · Modifications · Nanocomposites · Waste · Food packaging · Biopolymer

3.1 Introduction

Biopolymer-based nanocomposites have garnered tremendous attention in recent times. They are profoundly utilized in packaging industries, biomedical applications, energy storage, automotive industries, textiles as well as bio-sorbents for environmental remediation. These biopolymer-based nanocomposites are biodegradable, renewable, less toxic, and less expensive along with enhanced thermal, mechanical, and functional properties compared to conventional composites (Ates et al. 2020). Various natural biopolymers, such as starch, cellulose, lignin, chitosan, gelatines, alginates, hyaluronic acid, polyhydroxyalkanoates (PHAs), polyhydroxybutyrate (PHBs), and polyhydroxybutyrate-valerate (PHBVs) from plants, animals,

A. Suleman · J. Bora · A. J. Kalita · B. Chetia (✉)
Department of Chemistry, Dibrugarh University, Dibrugarh, Assam, India
e-mail: bolinchetia@dibru.ac.in

P. M. Visakh (ed.), *Engineering Applications of Polymer-Based Nanocomposites*, Advanced Structured Materials 237, https://doi.org/10.1007/978-3-031-93463-6_3

and microorganisms, and many more, have been used in the synthesis of such nanocomposites for various applications. Some synthetic polymers such as polycaprolactone (PCL), polylactic acids (PLAs), aromatic co-polyesters (e.g., PBAT), and aliphatic co-polyesters (e.g., PBSA) have also been used (Puiggalí and Katsarava 2017).

In recent times, thin films made up of biopolymer-based nanocomposites are widely being used in the packaging industry as a substitute for petroleum products. These films provide a sustainable solution to problems like multi-material packaging material and contamination with product remnants. These nanocomposite-based films also provide a reduced transmission rate of water vapor, protection from UV and visible radiation, safeguard food products from oxygen permeation, and control the loss of aroma. These bio-nanocomposites are biodegradable and thus solve the issue of disposal management reducing the overall environmental footprint (Pal et al. 2021) (Fig. 3.1).

Polysaccharides are widely used as biopolymers for the synthesis of nanocomposites to be used in food packaging. Starch, Chitosan, and Cellulose are the most widely employed polysaccharides, of which Starch is the most abundant, cheap, renewable, and non-toxic. However, starch on its own, lacks the required mechanical properties to be used as films for food packaging. Similarly, compounds based on starch are also very brittle, highly hydrophilic, and not very processable thus requiring surface chemical modifications by additive inclusion. Starch-based materials fortified with nano polymers such as nano clay, nanocellulose, chitin nanoparticles, and carbon-based materials are particularly used for applications in agriculture and packaging. Many nanomaterials such as TiO_2, ZnO, CuO, Ag, and Au nanoparticle reinforced starch composite exhibit antimicrobial properties making them suitable for food packaging

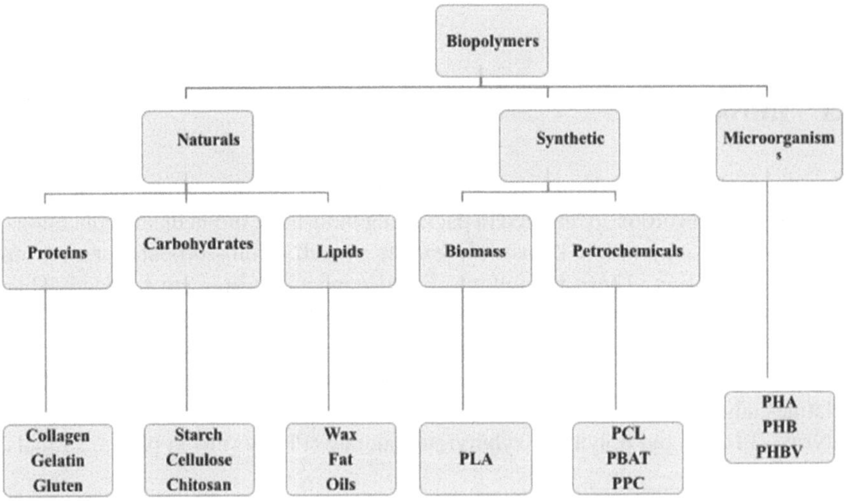

Fig. 3.1 Different categories of biopolymers based on source of origin

applications. The incorporation of nanofillers and additives having antioxidant and antimicrobial properties enhances the performance of starch-based nanocomposites while maintaining or minimally affecting their biodegradability (Lichade et al. 2021).

These improvements increase the shelf life of food items by preventing food spoilage and adulteration. Lifecycle assessments highlight the lower environmental impact of starch-based composites compared to traditional petrochemical-based packaging materials. As a sustainable alternative, these materials meet the demand for eco-friendly packaging solutions, reducing waste and reliance on non-renewable resources while promoting environmental conservation and sustainability in the food industry.

3.2 Starch

Starch, a polysaccharide with chemical formula $(C_6H_{10}O_5)_n$, is the primary food reserve in plants and also constitutes about 67% of carbohydrate intake in humans. Based on their digestibility starches are classified into three categories: rapidly digested, slowly digested, and resistant starch. Starch is made up of two major components: amylose and amylopectin, both of which are polymers of glucose. Amylose is a high branching polysaccharide consisting of linear polymer glucose units through alpha-1,4-glucose chains, whereas amylopectin is a high branching, almost unbranching polysaccharide with linear alpha-1,6-glycosidic linkages and alpha-1,4-glucose chains (Fig. 3.2). Amylopectin is the only component present in natural starch, which comprises about 75–90% of the relative dry weight of the starch, whereas amylose is a smaller constituent (Imberty et al. 1991).

Amylopectin's molecular structure plays a substantial role in determination of functional properties of starch. Depending on the chain length of Amylopectin, variations are observed in the physicochemical properties of starch. High side chain concentration of Amylopectin generates a double helix structure which leads to starch microcrystals. Hydrogen bonds adheres the amylopectin side chains, each consisting of about a dozen glucose units, resulting in the formation a double helix. Concentration of Amylose also plays an important role in the physicochemical properties of starch. High concentration of amylose leads to significant decrease in the average molecular weight, lamellar peak strength, degree of short-range order and crystallinity (Shen et al. 2018).

Starch naturally occurs as quasi-crystalline granules in the chloroplasts of leaves whereas it is commercially procured from seeds particularly corn, waxy corn, high amylose corn, wheat, rice and also from tubers and roots such as potato and sweet potato. Granules of natural starch are composed of crystalline layers known as the 'growth rings' and a number of alternating amorphous or semi-crystalline layers. The growth rings consist of two amylopectin helixes with systematically arranged chains, while the amorphous layer comprises of branched amylopectin and amylose. The major use of starch is in food whereas industrial application of starch involves

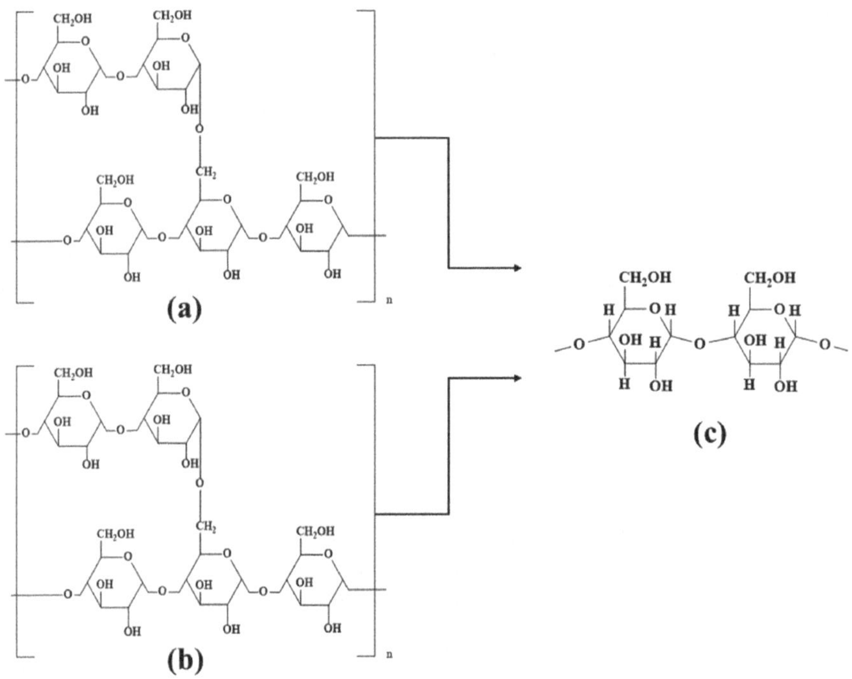

Fig. 3.2 **a** Structure of amylose, **b** structure of amylopectin, **c** structure of starch

sizing of paper and textiles, as adhesives and as packaging material in food packaging industries (Yang et al. 2017).

3.2.1 Modification Methods of Starch

Due to the incapability of naturally available starch to bear low acidity, shear force, or extreme temperatures throughout processing, it cannot be used in industrial applications. Numerous methods for the modification of starch can be utilized to get over these drawbacks. For the application of the starch in pharmaceuticals, paper, textiles, and food industries, it should be able to bear cold or hot temperatures, raised rates of stresses and shear along with weak acidity. These properties can be attained via genetic, enzymatic, chemical, and other physical techniques, such as freezing, annealing, oxidation, heat-moisture medication, ultrasonication, cross-linking, acid thinning, and hydroxypropylation. Modified starches enhance the stability of emulsions via the prevention of emulsion droplets from amalgamating via steric interactions. Additionally, modified starches can be utilized to develop food-grade emulsions affected lesser via salt, pH, and heat treatment, without surfactants.

(A) Physical Modification Methods

Some physical techniques such as application of moisture, heat, shears, or radiations are used to alter the properties of naturally occurring starch molecules. Since this method avoids the utilization of chemicals, it is beneficial for food products.

Heat moisture treatment (HMT), which is a kind of physical modification method, is a secure and eco-friendly hydrothermal refinement that adjusts moisture levels and temperature for the modification of starch. This method encompasses heating starch granules at a temperature ranging from 84 to 120° C with exposure to low water content (less than 35%, w/w) for a certain time period (0.25–16 h), lower than gelatinization temperature. Numerous physicochemical properties, such as time required for pasting, solubility, swelling ability, viscosity, crystallinity, endurance, and heat gelatinization, got altered. During the modification process, presence of lesser moisture content causes more perceptible surface changes of starch thus making it more susceptible to later enzymatic, physical, and chemical modifications. Starches which have endured HMT show a noticeably lesser swelling power at 25% moisture level. After enduring HMT modification, notable reduction in their peak viscosity of all starches was observed.

(i) Annealing

In the process of annealing, the starch undergoes heating treatment for a time of minimum 12 h by preserving it moist at a temperature above the glass transition point but below its gelatinization point. During the reconstruction of the structure of semi-crystalline starch granules, enhanced interactions with reduced free energy within the starch chains take place which results in the stability of the granules. In the process of formation of the starch slurry using water, the separated liquid state leads to the amylose leaching and alters the starch's composition.

As the annealing procedure is conducted at a lesser temperature than the starch's gelatinization temperature, it alters the chemical and physical properties of starch without causing any damage to the structure of the granules. Because of the reorganization of the structure during this process, the solubility and swelling capability of annealed starch are reduced. During this process of annealing, mobility occurs in both the crystalline and amorphous areas of starch, which leads to the creation and structuring of helical structures with contacts between amylopectin and amylose. Solubility, swelling power, and peak viscosity are reduced significantly in Bambara groundnut starch after annealing (Biliaderis 2009).

(ii) Ultrasonication

It is an environment friendly method for the transformation of native starch, since the modifications can alter its physicochemical properties, thus influencing the flow behavior of the starch suspensions. Sonocamped can be termed as a substitute of conventional methods like filtering, boiling, polymer depolymerization, and extracting.

In sonicated starch various alterations such as enhanced solubility, transparency, reduction in viscosity and disintegration of the granular structure is observed. In the sonication process, the parameters, such as time, power, frequency, and temperature along with other variables such as starch's source in plants, physical state, and concentration of the slurry and cumulatively these parameters, effect these transformations. This ultrasonication process offers numerous advantages, e.g., increased yield, shorter processing time, lowered chemical reagents and heat usage, lower pathogen content, improved quality of product and improved outcomes.

Starches originated from various sources such as corn, rice, wheat and potatoes demonstrated increased swelling strength and solubility after the process of ultrasonication. On exposure to numerous amalgamations of temperature, wave amplitude, duration, and starch content, improved paste clarity with reduced gelatinization enthalpy was shown by ultrasonicated maize starch. During the process of ultrasonication modification, granule depolymerization produced via cavitation induction which is an outcome of shear stress, is responsible for lowering the viscosity of the paste. Larger-granule starches are likely to be more responsive to ultrasonication and while decreasing other characteristics of starch, e.g., crystal structure, hardness, heat content values, adhesiveness, brittleness, recoverability, and elasticity, it can improve the paste's transparency.

While ultrasonication reduces the peak viscosity, it has no recognizable influence on the pattern of potato and millet paste starches when compared to native starch. Sonicating starches in water produced better results than sonicating them in ethanol. This increase in swelling power and solubility is attributed to both the H-bonding of free hydroxyl groups of amylose and amylopectin with water molecules and the damage done to the starch's crystalline molecular structure (Babu et al. 2019).

(B) Chemical Modification Methods

Molecular stability of Starch can also be enhanced by chemical modifications such as incorporation of suitable amounts of chemical compounds. This combination alters the viscosity, gelatinization as well as other properties of starch making it stable. The effect of these modifications is analyzed using a parameter called Degree of Substitution (DS) which refers to the mean number of hydroxyl groups replaced by the chemical modification in a single glucose unit.

(i) Phosphorylation/Cross-linking

Starch granules are strengthened upon cross-linking by preventing viscosity loss and granule breakage. This modification method involves a bifunctional reagent treatment enhancing the freeze-thaw resistance and degeneration resistance of starch. These modification methods utilize chemicals such as epichlorohydrin, adipic or citric acid, and phosphorus oxychloride which form a metaphosphate intermolecular cross-linking. This treatment prevents early gelatinization of starch as well as reduces the pace of retrogradation by reducing the movability of non-crystalline chains by forming intermolecular bridges.

By cross-linking phosphate groups with a maximal substitution degree, water solubility, paste clarity, gel consistency, shear resistance, and water absorption get improved. Several studies have shown that starch cross-linking reduces swelling of starch granules by improving intramolecular and intermolecular bonds (Tharanathan 2005).

(ii) Octenyl Succinic Anhydride

A typical esterification reaction is used to introduce hydrophobic Octenyl groups into hydrophilic starch granules using Octenyl succinic anhydride (OSA). The emulsion stability of OSA starches with higher molecular weight is better, while those with lower molecular weight have better encapsulation performance and efficiency. In the process of chemical modification, short linear amylose chains are released as a result of hydrolysis of the amylopectin chain ends. OSA-modified starches exhibit distinct digestive characteristics compared to their native forms, leading to the formation of slow-digesting or resistant starch (Sweedman et al. 2013).

(iii) Acid Thinning

Acid-alcohol modification involves heat treatment of starch at a temperature lower than 50 °C for a couple of hours in presence of acid. As a result, the glycosidic linkages between amylopectin and amylose are hydrolyzed. The crystalline part of the granules remains unaltered during the treatment process, and the reaction occurs in the non-crystalline areas, generating more crystalline starch. Consequently, the modified starch upholds its granular structure while displaying reduced viscosity and swelling potential, along with boosted gel strength and solubility. Acid-alcohol modified starch exhibits good film-forming capacity and is thus used in the production of consumable and biodegradable films for food packaging applications. Compared to native starch, the treated starch demonstrates enhanced solubility, boosted gel stability, and a higher degree of crystalline structure, while showcasing reduced viscosity and molecular weight (Tharanathan 2005).

(iv) Acetylation

Acetylation refers to the process of replacing hydroxyl groups on the starch chain with hydrophobic acetyl group using acetic anhydride. Based on this modification, starch can be classified into three categories: high, low, and medium DS starches. The electron cloud of acetyl group is comparatively far larger than the cloud of the hydroxyl group. Modified starch compounds thus experience steric hindrance even though the compounds attempt to stay in the stable low energy state. Acetyl groups present in the modified starch disrupts the hydrogen connection between the granules thus providing resistance again retrogradation (Shaikh et al. 2019).

(v) Hydroxypropylation

Hydroxypropylation process refers to etherification of native starch into non-ionic starch ether upon treatment with propylene oxide in the presence of an alkaline catalyst. Incorporation of hydrophilic hydroxypropyl group leads to weakening of hydrogen bonds causing weakening of the granules. This change leads to an increase in the rotational flexibility of the amorphous areas of the starch granules thus increasing its swellability. Starch modified using hydroxypropylation can thus be used as inactive ingredients in formulations for enhanced freeze-thaw stability and extended shelf life (Singh et al. 2007).

(vi) Oxidation

Oxidized starch is produced when native starch is treated with oxidizing agent such as sodium hypochlorite. The modified starch thus obtained is whiter as compared to native starch and has better potential to produce stable pastes. Oxidized starch also exhibits resistance against retrogradation and gelatinization (Singh et al. 2007).

(vii) Enzymatic Modifications

Hydrolyzed compounds like syrups and maltodextrins are obtained upon enzymatic treatment of starch. Small straight molecules of starch are generated due to hydrolysis of amylopectin by the action of debranching enzymes. Such modified materials exhibit better emulsification capability (Pratiwi et al. 2018). Enzymes like N-amylase and glucoamylase are used for the enzymatic action on starch to produce porous starch.

Solubility of the modified material gets enhanced due to the drainage of amylose during enzymatic hydrolysis. However, post enzyme action, an increase in amylose concentration is observed which after heat treatment and retrogradation produced dense crystallites having enhanced resistance against digestion. Enzyme modified starches are profoundly used in baking industries (Dura et al. 2014).

3.3 Starch-Based Nanocomposites

One of the widely accessible polysaccharides found in fruits, vegetables, cereals, etc., is starch. Starch has been identified chemically as a polymer of glucose molecules. Glycosidic bonds bind these glucose monomers together. Thermoplastic starch (TPS), also known as native starch, has poor mechanical qualities (brittleness and fragility), low thermal stability, hydrophilicity, high water vapor penetration, vulnerability to environmental conditions (humidity, ripping, plucking, etc.), and incompatibility with hydrophobic polymers. Therefore, to enhance the physical, mechanical, and barrier properties, starch is combined with other natural and synthetic polymers or integrated with different nanomaterials. Nanoparticles exhibit distinct physical, mechanical, optical, magnetic, electrical, and other properties compared to bulk materials. They also have a surface area/volume ratio. Therefore, bio-nanocomposites have emerged as a potentially useful material to improve mechanical and barrier properties (Othman et al. 2021).

A substantially higher degree of hydrogen bonding can be explained by the fact that starch is regarded as a semi-crystalline polymer. This makes the granules of starch insoluble in cold water. This is explained by preserving the compact structure of the starch granules. The starch granules' compact shape prevents water intrusion. However, heating the starch granule suspension encourages water molecules to be absorbed, which causes the starch granules' compact structures to expand (Cheng et al. 2021). Starch reinforced with nanofillers, such as carbon-based compounds, nanocellulose, chitin nanoparticles, and nano clay, are covered below.

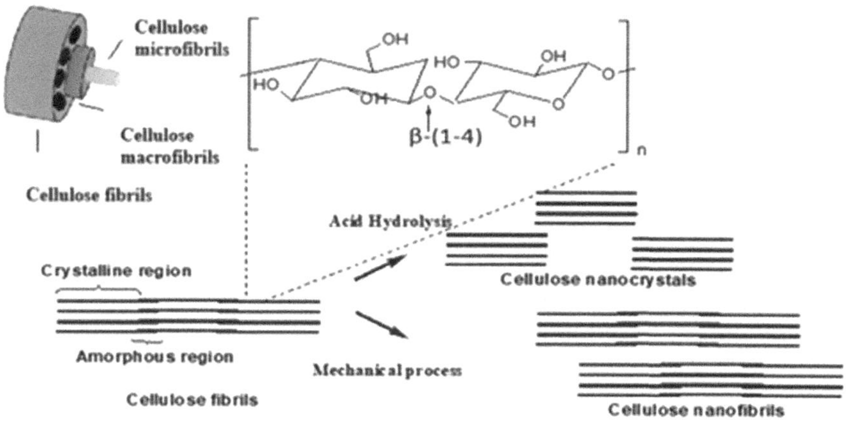

Fig. 3.3 Cellulose chemical structure and schematic diagram of the formation of cellulose nanocrystals and cellulose nanofibrils

3.3.1 Starch/Nanocellulose Composite

Cellulose is the main constituent of plant cell walls and can be derived from bacteria, fungi, algae, marine organisms, crustaceans, and plants. It is a polymer comprising D-glucose units connected by higher hydroxyl groups (-OH and -CH_2-OH) and (1,4)-glycosidic linkages. Higher stability is provided in equatorial positions (Fig. 3.3) (Othman et al. 2021).

Top-down and bottom-up methods can be used to produce nanocellulose. It can be produced in three different ways: (1) cellulose nanofibrils (CNFs), (2) cellulose nanocrystals (CNCs) derived from woods and other lignocellulosic materials using a top-down method, and (3) bacterial cellulose (BC) from bacterial biosynthesis using a bottom-to-top method.

Numerous industries, including biomedical engineering, the automotive industry, food packaging, cosmetics, electronics, construction, wood adhesives, textiles, and wastewater treatment, employ nano cellulose extensively. Starch and CNC nanocomposites were reported to decrease the water solubility (26.6–18.5%) and contact angle (38.2–96.3) while also improving the water-barrier characteristics, Young's modulus (112–520 MPa), and tensile strength (2.8–17.4 MPa) (Table 3.1).

3.3.2 Starch/Chitin Nanoparticles Composites

Next to cellulose, chitin is the second most prevalent natural polysaccharide. It is found in crustaceans' shells (crab, lobster, shrimp), arthropods' exoskeletons, squid's

Table 3.1 Types of nanocellulose

Cellulose nanofibrils or nanofibers (CNFs) nanofibrillated (NFC)/ microfibrillated (MFC) cellulose	Cellulose nanocrystals (CNCs)/ cellulose whiskers	Bacterial cellulose (BC) and electrospun cellulose nanofibers (ECNFs)
• The process from top to bottom is the breakdown of cellulose fibers into nanocellulose • Chemical treatments, mechanical treatments (such as steam expulsion, high-pressure homogenization, grinding, milling, and ultrasound), and a combination of these • Diameter: 5–50 nm, length: few micrometers	• Produced using a top-down process that breaks down cellulose fibers into nanocellulose • Acid hydrolysis • Diameter: 3–35 nm, length: 200–500 nm	• From bottom to top, bacteria or electrospinning produce low molecular weight carbohydrates or dissolved cellulose, respectively • Acetobacter xylinum and other bacteria are involved in biosynthesis • Pre-treatment is not necessary to eliminate lignin and hemicellulose • Long nanofiber system: 20–100 nm

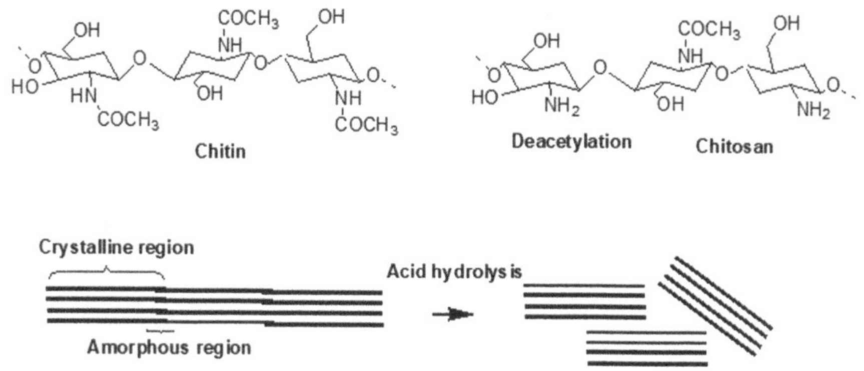

Fig. 3.4 Chemical structure of chitin and schematic diagram of the formation of chitin nanowhiskers

shells, mushrooms, algae, and fungi's cell walls. Chitin is made up of N-acetyl-2-amido-2-deoxy-β-D-glucose (N-acetylglucosamine) units connected by a β-(1,4)-glycosidic bond, where acetamide groups (-$NHCOCH_3$) are present at the C2 of the cellulose monomer (Heidari et al. 2021).

According to Cheng et al. Compared to chitin whiskers, chitin nanoparticles (CNPs) showed reduced crystallinity. Good interfacial interaction between CNPs' filler and starch matrix is achieved at low levels of CNPs by the plasticized potato starch/CNP nano composites tensile strength, glass transition temperature, storage modulus, and water vapor barrier characteristics (Cheng et al. 2021) (Fig. 3.4).

3.3.3 Starch/Nanoclay Nanocomposites

Clay is a 2D layered mineral silicate polymer compound. Two tetrahedral silicate sheets are positioned between an edge-linked octahedral sheet of magnesium oxide or aluminum oxide to produce a single layer. Automotive, aerospace, packaging, flame-retardant materials, therapeutic uses, and wastewater treatment are among the industries that use polymer/nanoclay composites. Smectite, chlorite, kaolinite, illite, and halloysite are among the kinds of nanoclays that can be distinguished. According to Mohan et al. when MMT nanoclay was added to corn starch-based film, there was a notable decrease in water absorption (by 22%), moisture uptake (40%), oxygen permeability (30%), and swelling thickness (31%), as compared to corn starch-based film. Additionally, the structure of the nanocomposite is determined by the concentration of MMT nanoclay. According to X-ray diffraction (XRD) investigation, the exfoliated structure forms in the polymer matrix at a lower concentration than the intercalated nanoclay structure, which forms at a higher concentration (>2%) (Mohan et al. 2017).

Halloysite clay nanotubes (HNTs), which are hollow cylinders of aluminosilicate, are a promising reinforcing material for polymers because they contain fewer hydroxyl groups on their surface than other silicates like MMT.

3.3.4 Starch/Carbonaceous Nanocomposites

Carbon allotropes commonly utilized in carbon-based nanocomposites include graphene, fullerenes, diamonds, and carbon nanotubes (CNTs).Single-walled carbon nanotubes (SWCNT) and multi-walled carbon nanotubes (MWCNT) are two types of CNTs that have been extensively researched as bolstering fillers for TPS nanocomposite films. CNTs have superior mechanical and thermal properties, a greater surface area, and electrical conductivity. They are extensively utilized in a variety of biological applications, environmental pollution management, sensing and detection, the automotive industry, and secondary food packaging. They also have better volume-to-area ratios than other nanoparticles (Famá et al. 1989) (Fig. 3.5).

There are currently few commercially viable starch-based products, though, because of significant disadvantages with their utilization, including their hydrophilic nature, poor processability, brittleness, and postprocessing crystallization. Numerous attempts to enhance starch-based materials by conventional composite formation produced somewhat better-quality materials, but their high viscosity hindered their processability.

But lately, there have been significant advancements in the mechanical, physical, and barrier qualities of starch-based nanocomposites that contain small amounts of fillers (Gamage et al. 2022) (Table 3.2).

Examples of starch nanocomposites supplemented with organic (cellulosic) fillers and inorganic clays or carbon nanotubes are included in this section.

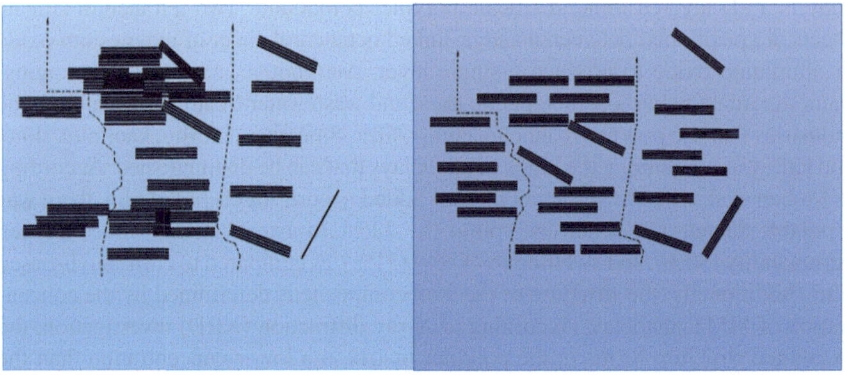

Fig. 3.5 Possible diffusion paths of gas molecules in **a** a traditional composite and **b** exfoliated polymer-clay nanocomposite. Note the increase in path length which ultimately results in decreased permeability (Dufresne et al. 2009)

Table 3.2 Summary of polymer nanocomposite trends as a function of the nature (crystalline or amorphous) of the matrix and its interaction with nanostructured fillers

Polymer matrix			
Property	Crystalline	Amorphous	Interaction
Elastic modulus	Increases w/volume fraction of filler Increases or no change w decrease in size	Increases w/volume fraction of filler Increases w decrease in size	Good
	Increases w/volume fraction of filler Increases w decrease in particle size Greater increase than for good interaction	Increases w/volume fraction of filler Increases w/decrease in particle size	Poor
Yield stress/ strain	Increases w/volume fraction of filler Increases w/decrease in particle size	N/A	Good
	Decreases w/addition of particlesDecreases w/ addition of particles	Decreases w/addition of particles	Poor
Ultimate stress/ strain	Increases w/decrease in particle size	Nano > micro after 20 wt.%	Good
	No unified result for change in volume fraction	Decrease w addition of particles	Poor
	Lower than pure for small volume fraction		
Density/volume	Increases volume w/ decrease in size NA	Increases volume w/ decrease in size	Good
		NA	Poor
Strain to failure	Decreases w/ addition of particles	Decreases w/addition of particles Increases w decrease in size	Good
	Decreases w addition of particles	Increases w addition of particles	Poor
Glass transition temperature	Decreases w addition of particles	Increases w/addition of particles	Good
	Decreases w/ addition of particles	Level until 0.5%, drops off level from 1–10 wt	Poor
Crystallinity	No major effect	N/A	Good
	No major effect	N/A	Poor
Viscoelasticity	Increases w volume fraction of filler Increases w/decrease in particle size	Increases w/addition of particles; drop at 1 wt.% flowed by increase in this property	Good
	N/A		Poor

From the above discussion, we can conclude that composites with inorganic fillers added to the starch matrix typically have better overall qualities. Numerous methods, including organophilization, silanization, alkylation, and surface etching, can alter the surface properties of fillers to improve particle dispersion, processability, and interface; this results in nanocomposites with superior qualities.

3.4 Food Packaging Applications of Starch-Based Nanocomposite

For the protection of food from spoilage and contamination via intrinsic and extrinsic factors, food packaging is a crucial medium. Starch is a readily available natural polymeric carbohydrate molecule which has great industrial applications. Because of their sustainability, starch-based nanocomposites show potential application as a promising food packaging material which enhance food preservation.

Starch comprises numerous excellent attributes such as large surface area, attractive self-assembly characteristics, and significant mechanical properties. Thus, it appears as an ideal candidate for the development of food packaging materials. Though starch plays a greater role in comparison to other traditional synthetic polymers, due to the insufficient mechanical and thermal characteristics in food packaging applications and strong hydrophilic nature of the starch films, its application is constrained. To enhance the thermal, mechanical, and barrier properties of starch-based polymers, utilization of nano-sized fillers is done. Along with starch, several food ingredients are utilized for the advancement of nanoparticles which consist of minerals, cellulose, lipids, protein, and surfactants (Dhull et al. 2020; Agarwal et al. 2020). These nanomaterials offer great biocompatibility and biodegradability in food packaging applications.

The change in composition of the nanoparticles occurs via their physicochemical, encapsulating, releasing, and protecting capabilities. To reduce this disadvantage, some new methods including nanoemulsions, nanocapsules, nanofibers, nanospheres, and nanoliposomes are introduced for the preparation of food packaging material (Guleria et al. 2020). Potato starch and rice starch nanocrystals are generally prepared via acid hydrolysis and they are commonly used in the starch-based films. Utilization of starch nanocrystals in the films has lowered water vapor permeability, lowered the surface fracture and roughness, and increased crystallinity. Due to the improved properties of starch-based nanocomposite packaging materials such as chemical stability, water vapor barrier properties, gas (carbon dioxide, oxygen) barrier properties, thermal stability, high mechanical strength, dimensional stability, recyclability, biodegradability, good optical clarity, heat resistance as well as developing active antimicrobial and antifungal surfaces, sensing and signaling biochemical and microbiological changes, they can be termed as a tremendous alternative for food packaging.

It is projected that the use of nanocomposite formulations will greatly extend the shelf life of numerous food types. Examples include packaging for cereals, processed meats, cheese, candy, and boil-in-bag foods; extrusion coating applications in conjunction with paperboard for dairy and fruit juice; and co-extrusion processes for the production of beer bottles and carbonated drinks.

The main drawbacks of using starch in food packaging applications are low barrier and mechanical properties along with sensitivity to moisture and UV-light. To overcome these drawbacks or to improve barrier and mechanical properties, starch has been incorporated with nanoparticles such as TiO_2, ZnO, graphene, and poly (methyl methacrylate-co-acrylamide). Starch sources considerably affect starch native granules. It makes them differ from crystalline and amorphous (9–10 nm), semi-crystalline and amorphous (100–400 nm), micro to nano range (<100 μm). Chemical modification processes such as hydroxylation and acetylation to replace the ether/ester group are mainly responsible to strengthen the mechanical properties of starch. Generally, starch-based films used for food packaging are potato and rice starch nanocrystals prepared via acid hydrolysis (3.16 M sulfuric acid) (Table 3.3).

In a study, it was reported that shoring of waxy maize starch nanocrystals in pea starch film improved tensile strength and diminished water vapor permeability, moisture content and water vapor transmission rate of the starch-based nanocomposite film significantly. Starch-based thermoplastic films have been widely utilized as food-containing casing/crates and food wrappers. Because of the poor strength, it has been adapted chemically or plasticized in order to remediate their strength (Pal et al. 2021; Martins et al. 2022; Adeyeye et al. 2023).

TiO_2 NPs supported on potato starch-based composite films showed great potential for food packaging since it led to a moisture uptake, lowering in water vapor

Table 3.3 Effect of nanostarch on the characteristics of packaging materials/bio-composite films (Adeyeye et al. 2023)

Sl. no	Type of starch crystal	Method employed	Properties improved
1	Mung bean starch	Acid hydrolysis	Reduced rate of water vapor transmission, water solubility while eruption strength of the food packaging material increased
2	Potato and rice starch	Acid hydrolysis	Reduced water vapor permeability, improved crystallinity, lowered the surface fracture and surface irregularity
3	Waxy maize starch	Acid hydrolysis	The tensile strength of the composite film enhanced, but water vapor transmission rate, water vapor permeability, and moisture content of the composite films significantly diminished
4	Rice starch	Acid hydrolysis	Tensile strength and crystallinity of the film enhanced further improving the water-barrier properties
5	Amadumbe starch	Acid hydrolysis	Thermal stability, opacity, tensile strength was enhanced whereas water vapor permeability of composite films decreased

permeability and water solubility and a raise of UV barrier properties along with tensile strength of the film. TiO_2 NPs and Na-MMT decrease the hydrophilicity and increase water vapor, UV barrier properties, and mechanical properties in cross-linked wheat starch. It makes them a perfect material for food packaging. Packaging film from starch/kefiran/ZnO NPs revealed enhanced UV barrier properties, tensile strength, thermal stability and Young's modulus which are essential for a food packaging system. Starch NPs/Ag NPs exhibited improved antibacterial activity against *Salmonella typhi*, *Staphylococcus aureus*, and *E. coli* and can be applied as a food coating material possessing antibacterial properties. Some starch-coated papers are highly used for food packaging applications. In order to create the coating substance for the paper, Jung and his group made a starch-based solution with Ag nanoparticles. The solution was made using a single-step synthesis method that included ultrasonication. After that, this was used to coat the paper for use in food packing (Pal et al. 2021). The paper's ability to withstand oil is enhanced by the coating. Additionally, the coated papers demonstrated sufficient antibacterial capabilities.

Another coating mixture was prepared via the combination of Ag nanoparticles and hydrophobic nanocrystalline porous starch. To enhance the mechanical and the oil resistance properties of the packaging paper, the mixture was applied to coat the paper. The coated paper possessed good antimicrobial properties against *Staphylococcus aureus* and *E. coli*. Besides, starch/ silver coatings, starch/silver nanocomposites were highly proposed in food packaging applications for their antibacterial properties (Pal et al. 2021; Mathew et al. 2019). Camu-Camu fruit was coated with starch/silver nanocomposite coatings to enhance the shelf life of the fruit by delaying the ripening of the fruits. The Ag nanoparticles were synthesized via green route were coalesced in the films at low concentrations in order that the prepared films did not exhibit any cytotoxic effects. The opacity and thickness of the starch films and conveyed antimicrobial activity increased via the incorporation of the nanoparticles. The Ag nanoparticles loaded starch-based films were capable of increasing the shelf life of fresh cheese samples up to 21 days. To develop Ag and ZnO-containing nanocomposite films, agar/cassava starch blends were used. Incorporation of the nanoparticles lowered the moisture content; water vapor permeability and water solubility of the agar/cassava starch blend films and a significant decrease in the development of *Staphylococcus aureus* and *Pseudomonas aeruginosa* was confirmed by the antimicrobial activity study. Ag-nanoparticle-loaded polyvinyl alcohol-blend film and boiled rice starch were also utilized in food packaging. After analysis, it was found that mechanical properties of the film increased over the only polyvinyl alcohol control film since the components of the films had physical interactions and the antimicrobial and optical properties of the films were also found to be better than the control film. These properties of the nanocomposite films were responsible for exploration in food packaging applications (Pal et al. 2021; Altaf et al. 2022).

Despite their enormous popularity in food packaging applications, starch-based films have poor mechanical and water resistance qualities. Nanocellulose (concentration range 0–10%) made from banana pseudostems was used to strengthen the native rice starch films in order to overcome these problems. By reducing the water vapor permeability, nanocellulose improved the film's mechanical stability. In an another modified starch-based film, i.e., glycerol plasticized starch films, cellulose nanofibers were utilized as the reinforcing agent (concentration of 0.4%, w/w) to improve the mechanical stability (Table 3.4).

Water vapor, strong heat stability, and oxygen barrier qualities are the main criteria used to choose a film for food packaging. A film with improved mechanical strength and pliability, guards against shock and other physical harm (Altaf et al.

Table 3.4 Summary of some starch-based nanocomposites utilized for food packaging (Pal et al. 2021)

Sl. no	Polysaccharide	Reinforcing agent	Antimicrobial activity	Food tested
1	Starch	Silver nanoparticle	–	–
2	Sugar palm starch	Silver nanoparticle	*E. coli, S. choleraesuis, S. aureus*	–
3	Hydrophobic noncrystalline porous starch	Silver nanoparticle	*S. aureus, E. coli*	–
4	Amazonian tuber starch	Silver nanoparticle	*S. aureus, E. coli*	–
5	Native rice starch	Nanocellulose extracted from banana pseudostems	–	–
6	Starch	Silver nanoparticle	*Salmonella spp., E. coli*	Fresh cheese
7	Thermoplastic starch	Cellulose nanofiber	–	–
8	Starch	Silver nanoparticles immobilized on cellulose nanofibrils	*S. aureus, E. coli*	
9	Plasticized sugar palm starch	Nanofibrillated celluloses from sugar palm fibers	–	–
10	Starch	Carrot cellulose nanofibers	*E. coli, L. monocytogenes,S. typhimurium, Penicillium spp.*	Grapes
11	Cationic starch	Nanoclay montmorillonite and ZnO nanopowders	–	–
12	Iranian potato starch	Montmorillonite	–	–
13	Cassava starch and polycaprolactone	Sodium montmorillonite	–	–
14	Thermoplastic starch	Montmorillonite	*Botrytis cinerea*	Strawberries
15	Starch	Montmorillonite	*E. coli*	–

2022). Linseed polyol enhanced the thermal stability, water absorption capacity, contact angle along with biodegradation of polyvinyl alcohol/corn starch film. When Ag NPs was introduced to it, antimicrobial behavior against *Candida albicans*, *Proteus mirabilis*, *E. coli*, *Staphylococcus aureus*, and *Klebsiella pneumoniae* among others was shown thus making it suitable for antimicrobial packaging applications. Improved chemical resistance, thermal stability, tensile strength, oxygen barrier qualities, and antibacterial activity against *Pseudomonas aeruginosa*, *E. coli*, *Staphylococcus aureus*, and *Bacillus subtilis* were all demonstrated by the poly (ethyl methacrylate)-co-starch/graphene oxide/Ag NPs (2 weight percent) composite film (Gamage et al. 2022) (Table 3.5).

Nanocellulose reinforced plasticized corn starch films with enhanced flexibility, mechanical strength, and oxygen and water vapor barrier properties, exhibits an advantageous effect on minimizing the oxidation of oil throughout the process of storage. This film demonstrated strong heat stability, which stops additional oxygen and water vapor transport. Furthermore, the storage investigation reveals that this starch-based nanocomposite can be utilized as an alternate packaging material for storage of edible oils for a period of length of approximately three months without altering the quantity of oil. When coupled with carboxyl methylcellulose (CMC)/Na-MMT, corn, potato, and wheat starch showed exceptional promise for use in food packaging. When combined with starch, sugarcane bagasse-derived cellulose nanocrystals (CNC) improved mechanical and water-barrier properties, decreased surface hydrophilicity, and increased water resistance, making this starch-based nanocomposite an excellent food packaging material. Another starch-based packaging material sweet potato starch (SPS)/MMT/thyme essential oil (TEO) showed that the introduction of MMT enhanced the water barrier and mechanical properties of SPS and reduced biodegradability. Nevertheless, integrating TEO reduced the elongation, tensile strength, water barrier, and Young's modulus with enhanced biodegradability in SPS/MMT. The microbial growth in pork meatballs stored under refrigeration and ambient conditions was significantly inhibited by the starch-based composite made from cassava starch, glycerol (2%), and cinnamon essential oil (2.5%). It demonstrated antibacterial properties against *Staphylococcus aureus*, *Salmonella typhimurium*, and *E. coli*. Potassium sorbate, a widely used preservative, was added to starch/nanoclay films to restrict sorbate migration and maintain antibacterial action for an extended length of time. Chen and his team combined starch/polyvinyl alcohol (PVA) with microfibrillated cellulose (MFC) and cinnamon aldehyde to create a controlled-release active starch-based film. With less flexibility, the addition of MFC was the cause of the increased hydrophobicity, crystallinity, tensile strength, and antibacterial activity (against S. putrefaciens). At 2.5 and 1.0% MFC, the water vapor and oxygen permeability decreased, but at higher concentrations, they improved. Additionally, MFC regulated the release of cinnamon aldehyde at 1 and 7.5%.

An active bilayer comprising a gelatin-polyvinyl alcohol-carbon dots (GPC) layer and a maize starch-polyvinyl alcohol-cinnamon essential oil (CPC) layer has been developed for food preservation utilizing 3D printing technology, as seen in Fig. 3.6. The film's powerful antioxidant and UV-blocking properties come from the carbon

Table 3.5 Features of the starch-based materials for applications in food packaging (Zubair et al. 2024)

Sl.no	Materials	Features packaging system of food	Strength features	Thermal characters	Applications
1	Cassava starch modified with boric acid	Water vapor permeability, solubility, and swelling capacity of the films decreased	Improved tensile strength	–	Food packaging
2	Corn starch/ polylactic acid bilayer film	Improved transparency and UV blocking properties, enhanced water absorption, oxygen, and water vapor permeability	Tensile strength and elongation at break (EB) enhanced	Improved thermal stability in comparison to the pure starch	Food packaging
3	Cinnamon essential oil and starch (CEO)-TiO_2 NPs	Enhanced oxygen barrier, water vapor permeability and mechanical strength, antibacterial action against S. aureus, S. typhimurium, and E. coli	Enhanced tensile strength	–	Food additive for packaging of raw pistachios
4	Starch—PBAT	Enhanced hydrophobicity, flexibility, mechanical properties, and moisture barrier	Tensile strength and elongation at break (EB) enhanced	–	Potential ingredient in food packaging
5	Corn starch-based films (curcumin-loaded bioactive nanocomplexes)	Enhanced gas barrier, antioxidant, and optical properties	Mechanical properties improved	Improved thermal stability	Packaging of blueberries

(continued)

Table 3.5 (continued)

Sl.no	Materials	Features packaging system of food	Strength features	Thermal characters	Applications
6	Cassava starch polyvinyl pyrrolidone-alizarin nanoparticles (PVP-AZ NPs)	Improved transparency and UV Blocking	Enhanced tensile strength	Significantly improved the stability	Packaging of shrimps
7	ZnO-rosemary polyphenols found in starch and PVA	Improved tensile toughness, strain and EB, decreased water vapor permeability, antioxidant activity, and antimicrobial properties against E. coli	Tensile strength and elongation at break (EB) enhanced	–	Active ingredient in food packaging

dots (CDs) generated from banana peels. When applied to the CPC film at a concentration of 3% (w/w), the cinnamon essential oil (CEO) produces the best integrity, the least amount of surface wettability, and the least amount of crystallinity (Fig. 3.6).

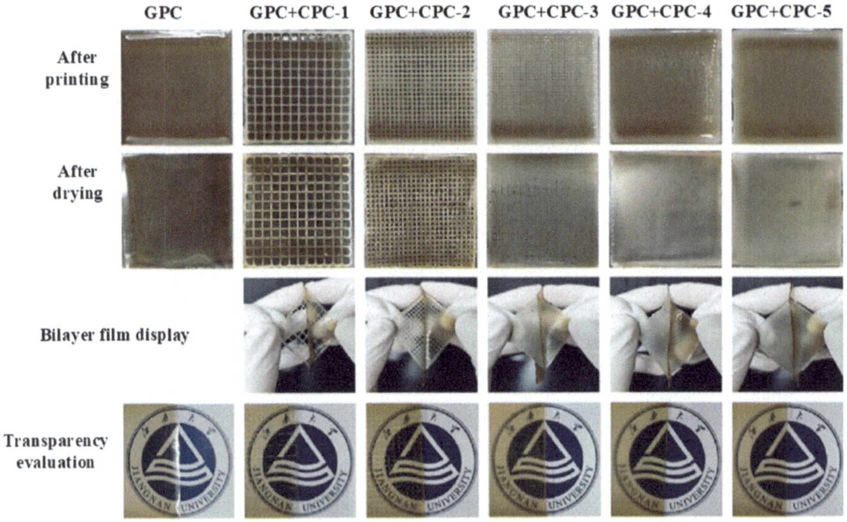

Fig. 3.6 Bilayer films derived from the GPC layer and a CPC layer after printing and drying and their transparency (Zubair et al. 2024)

Formulation of a bilayer coating prominently improved the stability and release efficiency of CEO. This process efficiently extended the shelf life of mangoes via preventing browning and rotting. Another study exposed the potential of creating a composite film with enhanced antibacterial and antioxidant properties via incorporation of carrot-derived CDS into corn starch. This film has been proved as a promising material for food preservation applications by efficiently preserving the quality of deep-fried meatballs. Packaging materials to monitor the spoilage of milk packed in a bottle were prepared via incorporation of pH indicators such as methyl orange (MO) and bromocresol green (BG) into a starch/nanoclay nanocomposite. For monitoring food quality, nanometals such as Ag_2O, TiO_2, MgO, SnO_2, CuO, and ZnO can be used in gas sensors (Gamage et al. 2022; Zubair et al. 2024).

3.5 Conclusion

Biodegradable and sustainable nanocomposites derived from starch address environmental concerns while ensuring food safety. Nanofillers enhance the mechanical strength, thermal stability, and barrier properties of these materials, making them suitable for food packaging. While challenges such as moisture sensitivity and large-scale production persist, ongoing research and technological advancements continue to improve their performance. As the world prioritizes sustainability, starch-based nanocomposites have the potential to transform food packaging, minimize plastic waste, and drive the adoption of environment friendly alternatives for a more sustainable future.

References

Adeyeye SAO, Babu AS, Guruprasath N, Ganesh PS (2023) Starch nanocrystal and its food packaging applications. Curr Res Nutr Food Sci 11:01

Agarwal A, Pathera AK, Kaushik R, Kumar N, Dhull SB, Arora S, Chawla P (2020) Succinylation of milk proteins: influence on micronutrient binding and functional indices. Trends Food Sci Technol 97:54

Altaf A, Usmani Z, Dar AH, Dash KK (2022) A comprehensive review of polysaccharide-based bionanocomposites for food packaging applications. Discover Food 2:10

Ates B, Koytepe S, Ulu A, Gurses C, Thakur VK (2020) Chemistry, structures, and advanced applications of nanocomposites from biorenewable resources. Chem Rev 120:9304

Babu AS, Mohan RJ, Parimalavalli R (2019) Effect of single and dual-modifications on stability and structural characteristics of foxtail millet starch. Food Chem 271:457

Biliaderis CG (2009) Structural transitions and related physical properties of starch. Starch 293–372

Cheng H, Chen L, McClements DJ, Yang T, Zhang Z, Ren F, Miao M, Tian Y, Jin Z (2021) Starch-based biodegradable packaging materials: a review of their preparation, characterization and diverse applications in the food industry. Trends Food Sci Technol 114:70

Dhull SB, Sandhu KS, Punia S, Kaur M, Chawla P, Malik A (2020) Functional, thermal and rheological behavior of fenugreek (*Trigonella foenum–graecum L.*) gums from different cultivars: a comparative study. Int J Biol Macromol 159:406

Dufresne A, Medeiros ES, Orts WJ (2009) Starch-based nanocomposites. In: Starch, characterization, properties and applications, pp 205–251
Dura A, Błaszczak W, Rosell CM (2014) Functionality of porous starch obtained by amylase or amyloglucosidase treatments. Carbohydr Polym 101:837
Famá L, Rojo PG, Bernal C, Goyanes S (1989) Biodegradable starch based nanocomposites with low water vapor permeability and high storage modulus. Carbohydr Polym 2012:87
Gamage A, Thiviya P, Mani S, Ponnusamy PG, Manamperi A, Evon P, Merah O, Madhujith T (2022) Environmental properties and applications of biodegradable starch-based nanocomposites. Polymers 14:4578
Guleria S, Kumar M, Kumari S, Kumar A (2020) Nano-starch films as effective antimicrobial packaging materials. In: Nanotechnological approaches in food microbiology. CRC Press, pp 437–453
Heidari M, Khomeiri M, Yousefi H, Rafieian M, Kashiri M (2021) Chitin nanofiber-based nanocomposites containing biodegrad able polymers for food packaging applications. J Consum Prot Food Saf 16:237
Imberty A, Buléon A, Tran V, Péerez S (1991) Recent advances in knowledge of starch structure. Starch-Stärke 43:375
Lichade KM, Jiang Y, Pan Y (2021) Hierarchical nano/micro-structured surfaces with high surface area/volume ratios. J Manuf Sci Eng 143:081002
Martins PC, Latorres JM, Martins VG (2022) Impact of starch nanocrystals on the physicochemical, thermal and structural characteristics of starch-based films. LWT 156:113041
Mathew S, Jayakumar A, Kumar VP, Mathew J, Radhakrishnan EK (2019) One-step synthesis of eco-friendly boiled rice starch blended polyvinyl alcohol bionanocomposite films decorated with in situ generated silver nanoparticles for food packaging purpose. Int J Biol Macromol 139:475
Mohan T, Devchand K, Kanny K (2017) Barrier and biodegradable properties of corn starch-derived biopolymer film filled with nanoclay fillers. J Plast Film Sheeting 33:309
Othman SH, Nordin N, Azman NAA, Tawakkal ISMA, Basha RK (2021) Effects of nanocellulose fiber and thymol on mechanical, thermal, and barrier properties of corn starch films. Int J Biol Macromol 183:1352
Pal K, Sarkar P, Anis A, Wiszumirska K, Jarzębski M (2021) Polysaccharide-based nanocomposites for food packaging applications. Materials 14:5549
Pratiwi M, Faridah DN, Lioe HN (2018) Structural changes to starch after acid hydrolysis, debranching, autoclaving-cooling cycles, and heat moisture treatment (HMT): a review. Starch-Stärke 70:1700028
Puiggalí J, Katsarava R (2017) Bionanocomposites. In: Clay-polymer nanocomposites, pp 239–272
Shaikh M, Haider S, Ali TM, Hasnain A (2019) Physical, thermal, mechanical and barrier properties of pearl millet starch films as affected by levels of acetylation and hydroxypropylation. Int J Biol Macromol 124:209
Shen X, Shang W, Strappe P, Chen L, Li X, Zhou Z, Blanchard C (2018) Manipulation of the internal structure of high amylose maize starch by high pressure treatment and its diverse influence on digestion. Food Hydrocoll 77:40
Singh J, Kaur L, McCarthy OJ (2007) Factors influencing the physico-chemical, morphological, thermal and rheological properties of some chemically modified starches for food applications—a review. Food Hydrocoll 21:1
Sweedman MC, Hasjim J, Tizzotti MJ, Schäfer C, Gilbert RG (2013) Effect of octenylsuccinic anhydride modification on β-amylolysis of starch. Carbohydr Polym 97:9
Tharanathan RN (2005) Starch-value addition by modification. Crit Rev Food Sci Nutr 45:371
Yang Z, Chaib S, Gu Q, Hemar Y (2017) Impact of pressure on physicochemical properties of starch dispersions. Food Hydrocoll 68:164–177
Zubair M, Rauf Z, Nawaz H, Shahzad S, Ullah A (2024) A review of recent advances in starch derived bionanocomposites for food packaging applications. Nano-Struct. Nano-Objects 39:101204

Chapter 4
High-Performance Polymer Nanocomposites' Based Solar Cells

Marissa Bailey Reichelscheimer, Saba Karimi, Homa Borna, Hande Bahar, Navid Nasajpour-Esfahani, and Hamid Garmestani

Abstract Polymer nanocomposites represent a transformative advancement in materials science, integrating nanoscale fillers into polymer matrices to achieve superior mechanical, thermal, and functional properties. These materials have gained significant attention for their applications in aerospace, automotive, biomedical, and renewable energy sectors. The classification of polymer nanocomposites is primarily based on the dimensionality of the fillers, including two-dimensional (2D), one-dimensional (1D), and three-dimensional (3D) nanostructures. Various fabrication techniques, such as melt blending, in situ polymerization, and solution processing, have been developed to optimize their dispersion and performance. A key focus of this chapter is the application of polymer nanocomposites in high-performance materials and emerging technologies such as solar cells, where they contribute to enhanced charge mobility, mechanical flexibility, and power conversion efficiency. Despite their advantages, challenges related to nanoparticle dispersion, interfacial bonding, and large-scale production remain critical barriers to widespread adoption. Addressing these challenges requires advancements in processing techniques, functionalization strategies, and environmental considerations.

4.1 Introduction

Polymers are reinforced with fillers of different sizes to enhance their mechanical, thermal, and physical properties. Traditional polymer composites have seen significant advancements with the incorporation of nanoscale fillers, leading to the development of polymer nanocomposites (Reynaud et al. 1999; Young et al. 2012). These materials provide superior strength, durability, and functionality, making them indispensable in various industries such as aerospace, automotive, medical, and renewable energy.

M. B. Reichelscheimer · S. Karimi · H. Borna · H. Bahar · N. Nasajpour-Esfahani (✉) · H. Garmestani
Department of Materials Science and Engineering, Georgia Institute of Technology, Atlanta, GA, USA
e-mail: Navid_nasajpour@yahoo.com

P. M. Visakh (ed.), *Engineering Applications of Polymer-Based Nanocomposites*, Advanced Structured Materials 237, https://doi.org/10.1007/978-3-031-93463-6_4

High-performance polymers (HPPs) are a subclass of polymers that offer exceptional stability and resistance to extreme environmental conditions. The integration of nanomaterials into HPPs has further expanded their applicability by creating high-performance polymer nanocomposites with enhanced properties.

4.2 Classification of Polymer Nanocomposites

Polymer nanocomposites are classified based on the size and shape of the dispersed nanoscale reinforcements.

4.2.1 Two-Dimensional (2D) Nanoscale Fillers

The first type of polymer nanocomposites involves two-dimensional (2D) nanoscale fillers such as layered silicates (Albdiry et al. 2013), graphene (Young et al. 2012; Du and Cheng 2012), or MXene sheets (Li et al. 2018; Tu et al. 2018), which typically have a thickness of just a few nanometers while their lateral dimensions can extend to hundreds or even thousands of nanometers. These fillers, when incorporated into polymer matrices, give rise to composites classified as layered polymer nanocomposites.

Sheet-Like Nanoscale Fillers

A key example of sheet-like nanoscale fillers frequently utilized in polymer nanocomposites is layered silicates, which belong to the 2:1 phyllosilicate family. These materials possess a crystal lattice structure composed of two-dimensional layers, with each layer measuring approximately 1 nm in thickness. The lateral dimensions of these layers range from a few tens of nanometers to several micrometers, and in some cases, even larger.

Polymer nanocomposites cannot be simply formed by physically mixing a polymer with layered silicates. This phenomenon is similar to that observed in certain polymer blends, where phase separation of layered silicates into distinct regions may occur. In immiscible systems, commonly found in conventional filled polymers, weak interactions between inorganic fillers and the polymer matrix result in inferior mechanical and physical properties. Conversely, strong interactions between layered silicates and polymers can facilitate nanoscale dispersion of inorganic phases within the polymer, leading to polymer nanocomposites that exhibit unique mechanical and physical characteristics distinct from those of micro-composites (Giannelis et al. 1999; Giannelis 1996; Okada et al. 1989; LeBaron et al. 1999; Vaia et al. 1999; Fauchais 1983).

The controlled synthesis of spherical nanoparticles remains a primary objective in many research efforts, as the size of these particles plays a crucial role in determining their properties. By incorporating nanoscale fillers into polymers, the advanced properties of these materials can be harnessed effectively. Nanofillers enable polymer matrices to attain characteristics that are otherwise unachievable with micro-sized fillers alone. For instance, adding nanoparticles in appropriate amounts can enhance both strength and modulus while preserving ductility, a balance that is often difficult to achieve using only micro-sized fillers.

The size of nanoparticles is primarily dictated by the production method employed. The thermal plasma method has been widely used for the fabrication of ultrafine powders (Fauchais 1983; Bogush et al. 1988; Young and Pfender 1985). Compared to the flame method, the plasma approach offers several advantages, such as reduced residence time of reactive species in the plasma reactor, rapid cooling of vaporized raw materials, and the ability to synthesize high-temperature boiling substances. Furthermore, the arc plasma technique yields non-aggregated particles, whereas the flame method tends to produce large particle aggregates. Additionally, uniform silica nanoparticles with a monodisperse size distribution can be synthesized via controlled hydrolysis of tetraethyl orthosilicate (TEOS) in ethanol, followed by condensation of dispersed phase material (Bogush et al. 1988).

Due to their extremely small size, nanoparticles have a high tendency to aggregate, making their uniform dispersion in polymers a challenging task. To ensure optimal performance in polymer nanocomposites, dispersants are often employed to enhance the distribution of nanoparticles within the polymer matrix.

4.2.2 One-Dimensional (1D) Nanoscale Fillers

The second category includes fillers with two dimensions at the nanoscale while the third dimension is considerably longer resulting in elongated structures. Examples of such one-dimensional nanoscale reinforcements are nanofibers and nanotubes such as carbon nanofibers, carbon nanotubes (Calvert 1999), and halloysite nanotubes (Liu et al. 2014). These fillers are widely used to enhance the mechanical and functional properties of composite materials.

4.2.3 Three-Dimensional (3D) Nanoscale Fillers

In contrast, the third type consists of nanocomposites that incorporate fillers where all three dimensions fall within the nanometer range. These materials typically involve low-aspect-ratio nanoparticles with near-spherical shapes such as silica nanoparticles (Reynaud et al. 1999; Young et al. 2012; Albdiry et al. 2013; Du and Cheng 2012; Li et al. 2018; Tu et al. 2018; Calvert 1999; Liu et al. 2014; Mark 1996), semiconductor nanoclusters (Herron and Thorn 1998), or quantum dots (Huang et al.

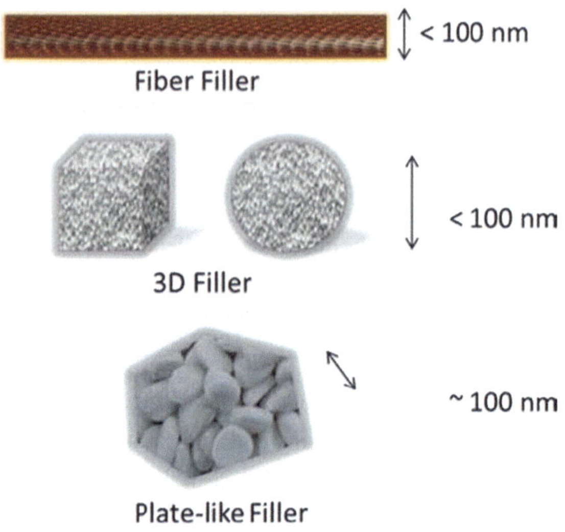

Fig. 4.1 Visual comparison of various types of nanoscale fillers illustrating their distinct shapes and dimensions (Khan et al. 2016)

2016). Figure 4.1 presents a visual comparison of various types of nanoscale fillers illustrating their distinct shapes and dimensions.

Polymer nanocomposites exhibit superior mechanical and physical properties compared to their base polymers, largely due to the extensive interfacial area formed between the polymer matrix and nanoscale fillers. This significantly large interfacial region, combined with the nanoscopic dimensions of the fillers, sets polymer nanocomposites apart from traditional composites. The predominance of interfacial regions arising from nanoscale phase dimensions has led to exceptional mechanical and physical behaviors that are not observed in conventional polymer composites. To further refine these properties, nanofillers are often functionalized to modify interfacial interactions and enhance the overall performance of polymer nanocomposites (Sahoo et al. 2010). The effectiveness of these fillers depends on their dispersion and interaction with the polymer matrix, which directly impacts the overall performance of the nanocomposite.

4.3 Preparation of Polymer Nanocomposites

The synthesis of polymer nanocomposites plays a crucial role in determining their structural integrity, dispersion quality, and overall material performance. By placing nano-filling material in a matrix of a general polymer, depending on the nanofiller chosen, the resulting composite will have specified properties (Park et al. 2002). Various processing techniques have been developed to ensure uniform nanoparticle dispersion, strong interfacial bonding, and optimized properties.

The selection of an appropriate processing method is crucial for obtaining high-quality polymer nanocomposites with superior performance characteristics. Several

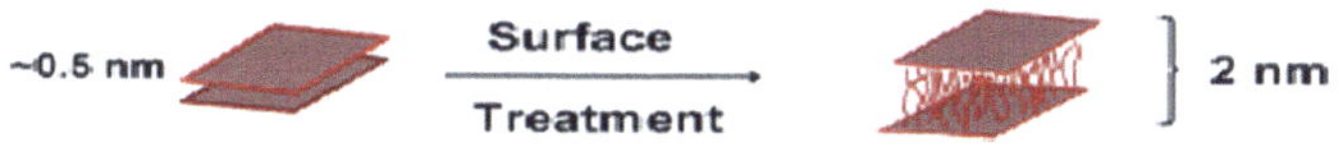

Fig. 4.2 Schematic showing of MMT surface treatment (Fu et al. 2019)

processing techniques have been developed for the fabrication of polymer nanocomposites (Ray and Okamoto 2003; Oriakhi 1998; Ma et al. 2003; Carrado and Xu 1998; Chen et al. 2003; Zheng and Wong 2003; Crosby and Lee 2007; Sithole et al. 2017; Coleman et al. 2006; Oliveira and Beatrice 2018; Uddin 2008), including melt mixing, solution mixing, in situ polymerization, and in situ particle processing, among others. However, a single universal approach for synthesizing all types of polymer nanocomposites is unattainable due to the diverse physical and chemical properties of polymers and the various available processing technologies. Each polymer system requires specific processing conditions tailored to the efficiency of the method, the type of nanoscale filler used, and the targeted properties. Consequently, different processing techniques often yield varying results (Park et al. 2002) (Fig. 4.2).

4.3.1 Layered Filler/Polymer Nanocomposites

Due to their hydrophilic nature, layered silicate clays do not naturally mix or interact well with most polymers. Additionally, electrostatic forces cause the clay platelets to stack tightly together, as adjacent platelets share counter-ions, forming compact stacks. The direct use of untreated clay in polymer nanocomposite production is ineffective because most platelets remain aggregated and do not interact efficiently with polymer matrices. To address this challenge, the surface of the clay must be modified before use. A common and effective method for surface modification involves ion exchange, which enhances compatibility with organic polymers. This treatment enables better integration of clay into polymer matrices, facilitating the separation of platelets so they can be more effectively intercalated and exfoliated. Depending on the specific composition of the layered silicate, organic cation, and polymer matrix, as well as the preparation technique used, three main types of polymer nanocomposites can be achieved when layered silicates are incorporated, as illustrated in Fig. 4.3 (Alexandre and Dubois 2000).

4.3.2 Methods for Producing Layered Silicate/Polymer Nanocomposites

The fabrication of layered silicate/polymer nanocomposites generally follows four main approaches: (1) polymer or pre-polymer intercalation from solution, (2) in situ

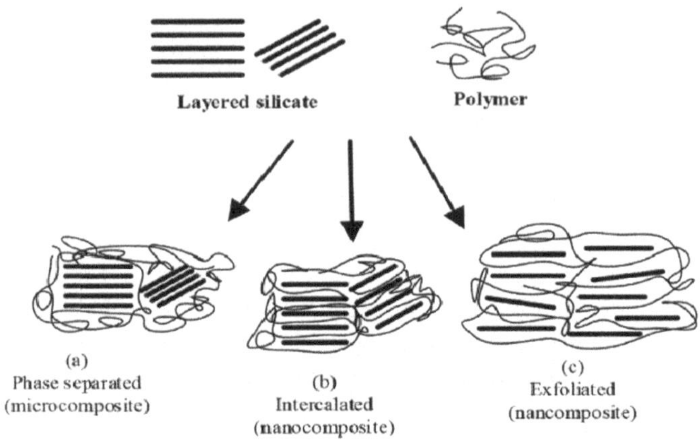

Fig. 4.3 Schematic illustration of three different types of nanocomposites arising from the interaction of layered silicates and polymers: **a** phase separated polymer micro-composite; **b** intercalated polymer nanocomposite; and **c** exfoliated polymer nanocomposite

intercalative polymerization, (3) direct melt intercalation, and (4) template synthesis (Alexandre and Dubois 2000; Ray and Okamoto 2003; Oriakhi 1998).

- In the first approach, the polymer or pre-polymer must be soluble, and the silicate layers must be expandable. Initially, layered silicates are swelled in a solvent such as toluene, chloroform, or water. After mixing with a polymer solution, polymer chains replace the solvent within the silicate interlayers, preserving the intercalated structure as the solvent evaporates.
- The second method, known as in situ intercalative polymerization, involves swelling the layered silicate in a monomer or monomer solution, allowing polymerization to occur between the intercalated layers. This technique gained significant attention following research on MMT/Nylon 6 nanocomposites by the Toyota research group (Young and Pfender 1985). A two-step approach has also been used, where treated MMT solutions are first prepared and subsequently blended with polymers (Ma et al. 2003).
- The third approach, direct melt intercalation, involves blending layered silicates with molten polymers, eliminating the need for solvents.
- Additionally, in situ hydrothermal crystallization enables clay layers to form within an aqueous polymer gel, leading to the synthesis of silicate-based nanocomposites. This process is particularly effective for water-soluble polymers such as hydroxypropyl methylcellulose (HPMC), poly(vinyl-pyrrolidone) (PVPyr), poly(dimethyldi-allylammonium) (PDDA), and poly(aniline) (PANI) (Carrado and Xu 1998). In situ polymerization is also applicable for the preparation of layered graphite-polymer nanocomposites (Chen et al. 2003; Zheng and Wong 2003).

Each of these fabrication techniques provides unique benefits depending on the polymer type, processing conditions, and targeted application. Also, a consideration that should be made when examining the processing techniques of polymer nanocomposites is the ability for these methods to be incorporated on a much larger scale. In that, it should be considered whether the resulting polymer nanocomposite can be produced repeatably, using a common solvent, and to a high yield degree. If these criteria are not met, the resulting polymer nanocomposite may not be able to be produced to a large degree that would be necessary for cost-effective industrial use.

4.4 High-Performance Polymers (HPPs)

In general, polymers are known for having adjustable properties that can adapt to changing environmental conditions, including a change in the external pH, ionic concentration, application of electromagnetic radiation, or temperature. Polymers are also versatile in that they can be developed in different phases. Specifically, they can form gels, solids, or liquids of varying rheological properties for a wide variety of applications. This makes polymers incredibly useful in sensing applications, where the chemical and physical nature of the polymer chain can be tuned to allow for the detection of a specific component, for example (Crosby and Lee 2007; Sithole et al. 2017).

High-performance polymers (HPPs) are engineered to provide exceptional thermal stability, mechanical resilience, and resistance to extreme environmental conditions. Unlike standard polymers, which degrade under heat, chemical exposure, or mechanical stress, HPPs maintain their integrity under rigorous conditions, making them indispensable for high-tech industries. Examples of High-Performance Polymers:

- **Polyether Ether Ketone (PEEK)**—Known for its superior thermal resistance and mechanical properties, widely used in aerospace, medical, and industrial applications.
- **Polyimides (PIs)**—Feature excellent electrical insulation, high-temperature stability, and chemical resistance, making them ideal for microelectronics and flexible circuits.
- **Polysulfones (PSUs)**—Possess strong oxidation resistance and hydrolytic stability, commonly used in medical and food-processing applications.
- **Polyphenylene Sulfide (PPS)**—Noted for its flame resistance and chemical stability, frequently employed in automotive and industrial components.
- **Liquid Crystalline Polymers (LCPs)**—Have remarkable tensile strength, making them suitable for lightweight, high-performance engineering applications.

The robust nature of these polymers stems from highly ordered molecular structures and strong intermolecular bonding, which grant them resilience under extreme conditions. Advanced polymerization techniques, including step-growth polymerization, ring-opening polymerization, and aromatic substitution reactions, have further enhanced their properties, allowing for greater versatility and industrial applications.

4.4.1 High-Performance Polymer Nanocomposites

The integration of nanomaterials into high-performance polymers has resulted in the development of nanocomposites that surpass the properties of conventional polymer composites. These materials exhibit unmatched mechanical, thermal, and electrical capabilities, making them highly desirable for industries that demand exceptional performance. Nanocomposites are made possible through the use of nanotechnological processes that combine the properties of different components in such a way that the final product can be tuned to have optimal characteristics.

Nanomaterials are characterized by their size; the fact that particles of a size on the order of a nanometer exhibit a greater surface area-to-volume ratio can allow them to be extremely influential in polymer matrix incorporation. Nanoparticles consist of material on the order of several nanometers that are added to the greater polymer matrix to adjust the overall properties of the resulting composite. Common Nanomaterials in Polymer Nanocomposites:

- **Carbon Nanotubes (CNTs)**—They offer superior tensile strength, thermal conductivity, and electrical properties. Carbon nanotubes have been implemented in applications including sensors, devices, catalyst materials, and energy storage materials (Coleman et al. 2006). The basic nature of the carbon nanotube is the single-walled carbon nanotube, consisting of a single layer of carbon arranged into a ring-like structure. A single-walled carbon nanotube is essentially a single exfoliated sheet of graphene bonded together to form a cylindrical tube. A current challenge inhibiting the widespread use of carbon nanotubes in global applications is that they cannot be easily and effectively produced on a large scale at this point in time. Current carbon nanotubes that have been produced for widescale application are considered to be impure, requiring more processing to increase the degree of purity. This can make the overall processing technique of forming carbon nanotubes incredibly costly, preventing their use in industrial applications.
- **Graphene and Graphene Oxide**—They enhance charge mobility, mechanical reinforcement, and structural integrity.
- **Metal Oxide Nanoparticles (TiO_2, SiO_2, ZnO)**—Provide UV resistance, antibacterial activity, and enhanced optical clarity.
- **Nanoclays**—Improve barrier properties, flame resistance, and mechanical strength. Clay/Nylon 6 has been researched by the Toyota company, proving to be useful in future applications (Oliveira and Beatrice 2018). Nanoclay materials are currently the most commercially used nanocomposite (Uddin 2008). Nanoclays are known for their lowered permeability of gaseous components and coefficient of thermal expansion, proving to be useful in applications where gaseous elements

or temperature are controlling environmental factors (Alexandre and Dubois 2000; Ray and Okamoto 2003; Paul and Robeson 2008). The natural state of nanoclay materials is in a layered structure that should be broken down to allow for the incorporation of these particles into a polymeric matrix to form a nanocomposite (Smart and Moore 1995).

There is a relationship between graphite, graphene, and carbon nanotubes, as these are all essentially composed of carbon on an elemental basis. Graphite can be considered as a plethora of carbon atoms connected by a series of covalent bonds to form a net-like structure. Graphene is a monolayer of graphite, or a layer of single atomic thickness exfoliated from the surface of graphite. The carbon nanotube is produced by bonding the sheet of graphene with itself, forming a tube. Multiple sheets of graphene can be incorporated, forming what is known as a multi-walled carbon nanotube.

Nanowires are similar to nanotubes in their nanoscale diameter size, but they are much longer in length. The length of a nanotube can be on the order of a micrometer or even a millimeter. For nanofibers composed of carbon, the diameter of the tube is usually bigger than that of a carbon nanotube and can be considered in different structures (Thostenson et al. 2005). As per the structure–processing–property relationship typically referenced in materials science and engineering, incorporating new atomic structure arrangements (through choice of a processing technique) has a significant impact on the properties experienced by that material.

Nanofiller material can be further characterized by their composition and dimension. One-dimensional nanofillers include nanotubular materials, two-dimensional nanofillers include the carbonic material graphene, and three-dimensional materials can be considered to be spherical in shape. The dimension associated with the nanofilling material used can have a significant impact on the properties of the composite, which influences the potential for use in further applications that have yet to be explored.

4.4.2 Fabrication Techniques for High-Performance Polymer Nanocomposites

To ensure uniform dispersion and strong interfacial bonding, various processing techniques are used:

- **In Situ Polymerization**—Ensures uniform dispersion of nanomaterials by integrating them into monomers before polymerization. This technique of polymerization is commonly used when the monomer of interest is a liquid, and the nanoparticle is soluble within it. In cases where the polymer matrix monomeric material and the nanofiller material are hydrophilic, techniques like in situ polymerization can be used. However, there are many cases where the monomer comprising the polymer matrix and the nanofiller material are not conducive to water interaction.

In the case of a material like a nanoclay and nylon composite, nanoclay has a surface layer of ionic components that are attracted to water. The clay should therefore be coated on the surface with an agent to make it hydrophobic so that it can interact to join a hydrophobic polymer matrix. This can be done by applying a surfactant.

- **Melt Blending**—A cost-effective industrial method that disperses nanomaterials into molten polymers. This technique of polymerization is used when a polymer and its suggested filler material are not soluble in one another or when a large degree of a polymer nanocomposite is to be produced. This involves the use of equipment including twin screw extruders to properly disperse the immiscible filler in the polymer matrix by applying high temperatures and high shear rates to the mixture.
- **Solution Processing**—Provides precise molecular-level dispersion, often used in electronics and coatings. This technique of polymerization is commonly used when both the polymer and the nanofiller material are soluble in a particular solvent. Solution processing requires that the nanocomposite that is produced in solution be dried and filtered or precipitated out to retrieve the solid, so this method may not be useful if the resulting nanocomposite material is soluble in the solvent in which the components are placed. Solution processing is also known to be useful in smaller applications, as it can become costly in resources, energy, and time.
- **Electrospinning**—Produces high-performance nanofiber-reinforced composites with enhanced mechanical and thermal properties. These techniques have enabled the production of polymer nanocomposites for applications such as automotive components, medical implants, high-strength coatings, energy storage devices, and flexible electronics. However, challenges such as nanomaterial dispersion, interfacial bonding, and large-scale production feasibility remain topics of active research.

Each of these techniques has its advantages and limitations, and the choice depends on desired properties and end-use applications.

4.5 Polymer Nanocomposites in Solar Cells

Polymer nanocomposites are playing a pivotal role in advancing solar energy technologies, particularly in organic photovoltaics (OPVs) and perovskite solar cells. Traditional polymer-based solar cells face limitations in charge mobility and efficiency, but the introduction of nanomaterials has significantly improved their performance.

4.5.1 Key Nanocomposite Materials for Solar Cells

- **Conjugated Polymer Blends**—Conjugated polymer blends, particularly those combining Poly(3-hexylthiophene) (P3HT) and (Tu et al. 2018; Tu et al. 2018)-Phenyl-C61-butyric acid methyl ester (PCBM), have been extensively researched for their application in organic photovoltaics (OPVs). These materials form the active layer in bulk heterojunction (BHJ) solar cells, where their complementary properties facilitate efficient charge separation and transport.
- Key Properties and Performance:
  - **P3HT**: A semiconducting polymer known for its strong absorption in the visible spectrum and good hole mobility.
  - **PCBM**: A fullerene derivative that acts as an electron acceptor, offering high electron mobility and favorable energy levels for charge transfer.

When blended, P3HT and PCBM create an interpenetrating network that enhances the donor–acceptor interface, crucial for effective exciton dissociation. Optimizing the morphology of this blend through processing techniques like thermal annealing and solvent selection has led to power conversion efficiencies (PCEs) exceeding 5% in some studies. These findings underscore the potential of P3HT:PCBM blends in advancing the efficiency and viability of organic solar cells (Saini 2015).

- **Graphene-Based Nanocomposites**—They have emerged as promising materials in the field of solar cells, offering significant enhancements in charge transport efficiency and device stability. By integrating graphene into various components of solar cells, researchers have achieved notable improvements in performance metrics.
- Key Properties and Performance:
  - Enhanced Charge Transport: Incorporating graphene into electron transport layers, such as TiO_2/graphene (TiO_2/G) composites, facilitates more efficient electron transfer from the perovskite layer. This enhancement is attributed to graphene's superior electrical conductivity, which reduces charge recombination losses and improves overall charge mobility within the device.
  - Improved Power Conversion Efficiency (PCE): The integration of graphene-based nanocomposites into perovskite solar cells has led to significant increases in PCE. For instance, perovskite solar cells utilizing TiO_2/G electron transport layers have demonstrated higher efficiencies compared to those without graphene, highlighting graphene's role in boosting device performance.
  - Stability Enhancement: Graphene's robust mechanical and chemical properties contribute to the long-term stability of solar cells. Devices incorporating graphene-based nanocomposites exhibit enhanced resistance to environmental degradation factors, such as moisture and thermal stress, thereby extending the operational lifespan of the solar cells (Ali 2023).

- **Quantum Dot-Polymer Hybrids**—They have garnered significant attention in solar cell research due to their tunable bandgap properties, which enable enhanced light absorption across a broad spectrum. By adjusting the size and composition of quantum dots (QDs), it's possible to tailor their electronic and optical properties, thereby optimizing solar cell performance.

- Key Properties:
  - Tunable Bandgap: The ability to modify the bandgap of QDs allows for the absorption of specific wavelengths of light, enhancing the overall efficiency of solar cells.
  - Enhanced Light Absorption: Incorporating QDs into polymer matrices can lead to improved light-harvesting capabilities, as QDs can be engineered to absorb light in regions where the polymer may be less effective.
  - Charge Transport Efficiency: The integration of QDs with conductive polymers facilitates efficient charge separation and transport, reducing recombination losses and improving power conversion efficiencies (Song et al. 2017).
- **Titanium Dioxide (TiO_2)** is extensively utilized in various solar cell technologies due to its favorable properties. Below are its key application:
- Dye-Sensitized Solar Cells (DSSCs):
  - Nanostructured TiO_2 Photoanodes: The architecture of TiO_2 nanostructures, such as nanoparticles, nanowires, and nanotubes, significantly influences the efficiency of DSSCs. These structures affect the photoanode properties, impacting light absorption and electron transport (Gnida et al. 2021).
  - Large-Diameter TiO_2 Nanotubes: Utilizing TiO_2 nanotubes with diameters around 500 nm allows for the integration of hierarchical and blocking layers. This design enhances electron transport and reduces recombination, leading to improved DSSC efficiency (Elzarka, et al. 2005).
- Perovskite Solar Cells:
  - Aerogel-Like TiO_2 Films: Innovative fabrication of ultra-porous TiO_2 films with aerogel-like properties has been achieved through plasma deposition. These films serve as effective electron-conducting electrodes in perovskite solar cells, demonstrating efficiencies comparable to traditional mesoporous anatase electrodes (Obrero et al. 2024).
- Advancements in TiO_2 Nanomaterials:
  - Black TiO_2 Nanomaterials: Recent studies have focused on black TiO_2 (B-TiO_2) nanomaterials, which exhibit enhanced light absorption and charge transport properties. These advancements contribute to improved performance in solar cell applications (Selema 2024).
- **Zinc Oxide (ZnO)**—ZnO is widely utilized in organic and hybrid solar cells due to its excellent electrical and optical properties. It serves as an electron transport layer (ETL) and enhances charge collection efficiency by reducing charge recombination. Properties and Advantages of ZnO in Solar Cells:

Fig. 4.4 Photograph of flexible plastic solar cell of conjugated polymer hybridized with ZnO nanorods. Circular patterns of silver are top electrodes (Tong et al. 2012)

- High Electron Mobility—ZnO nanorods and thin films exhibit electron mobilities ranging from 10^{-3} to 10^{2} cm^2/Vs, significantly improving charge transport compared to organic materials.
- Wide Bandgap (~3.37 eV)—Enhances transparency and allows efficient UV absorption without interfering with visible light absorption in active layers.
- Low Recombination Rates—ZnO nanostructures, such as nanorods and nanoparticles, facilitate efficient electron extraction and minimize carrier losses.
- Compatibility with Flexible Substrates—ZnO can be synthesized at low temperatures (~85 °C) using solution-based methods, making it suitable for flexible and roll-to-roll solar cell fabrication.
- Enhanced Power Conversion Efficiency (PCE)—Studies have demonstrated that ZnO-based hybrid solar cells can achieve efficiencies ranging from 0.2 to 2.7%, with optimal device structures improving performance further (Tong et al. 2012) (Fig. 4.4).

4.5.2 Advantages of Polymer Nanocomposites in Solar Applications

- **Higher Power Conversion Efficiency**—Enhanced charge separation and transport properties.
- **Improved Mechanical and Thermal Stability**—Nanomaterials help prevent structural degradation.
- **Lightweight and Flexible**—Compared to conventional silicon-based solar cells.

However, challenges remain, including:

- **Degradation Under UV Exposure**—Long-term stability issues.
- **Scalability**—Large-scale manufacturing is still complex.
- **Cost-Effectiveness**—High-quality nanomaterials are expensive.

Future research focuses on developing more stable and cost-effective nanocomposite-based solar materials for commercial adoption.

4.6 Challenges and Future Prospects of Polymer Nanocomposites

4.6.1 Challenges

Despite their advantages, polymer nanocomposites face several challenges that must be addressed before they can achieve widespread industrial adoption.

Nanoparticle Dispersion

Agglomeration of nanoparticles negatively impacts composite properties and performance. Achieving uniform dispersion remains a critical challenge.

Interfacial Bonding

Stronger adhesion between nanofillers and polymer matrices is needed for long-term durability. Functionalization techniques can help improve interfacial compatibility.

Processing Complexities

High melt viscosity and filler agglomeration affect industrial feasibility. Some polymer matrices degrade at high temperatures, limiting processing techniques. Exfoliation techniques help but require high shear forces that may damage polymer chains.

Standardization and Quality Control in Nanocomposites

Additional techniques should be standardized across global industries to evaluate the degree to which a polymer nanocomposite is of sufficient quality and pure. There is currently no standardized way to determine how effective a laboratory or manufacturer is at producing nanocomposites, which hinders their use in industrial applications (Kingston 2007). It is currently being investigated to use thermogravimetric analysis as a characterization technique to determine the degree to which a polymer nanocomposite is pure. Thermogravimetric analysis is a technique that measures the weight loss of a material as it is exposed to heat. In the case where the oxidation temperature of the polymer nanocomposite is higher and there is a greater mass of material remaining after exposure to heat, the material is said to be purer. Additionally, microscopic methods can be used to analyze the surface characteristics of a polymer nanocomposite and thus determine how well-integrated the filler material is. The US National Institute of Standards and Technology has proposed the use of thermogravimetric analysis for carbon nanotube materials and microscopy for single-walled carbon nanotubes (Arepalli et al. 2004).

Mechanical and Processing Limitations

Polymer nanocomposites are known for exhibiting a lower degree of toughness when compared to traditional polymer counterparts, significantly inhibiting the potential for their use. Additionally, it should be noted that these composite materials also have hindered optical properties and a higher melt viscosity. These two factors, especially the latter, can make it challenging to process polymer nanocomposites into particular shapes for various applications.

Environmental and Safety Concerns

Another consideration that must be made before polymer nanocomposites can be incorporated into more industrial applications is the safety of the material. Researchers have yet to determine whether or not a majority of the recent nanocomposite materials is safe for human exposure, and the production and incorporation of these materials result in exposure (which may be unsafe at this time). Some countries, including the UK and France have begun to incorporate standard policies outlining the unknown dangers of nanomaterials (HSE 2013, INRS 2012).

4.6.2 Future Research Directions

Several key research areas aim to overcome these challenges and expand the applications of polymer nanocomposites:

- **Advanced Functionalization Techniques**—To optimize nanoparticle dispersion.
- **Sustainable and Biodegradable Nanocomposites**—For greener applications.

It is also being investigated whether or not nanocomposite materials can be reused for multiple applications or recycled into new materials, both of which would allow for the increase in the sustainability of these materials. Some pre-existing policies could be expanded to include up-and-coming nanomaterials. For example, the European Union currently requires that electronics be recycled or reused instead of simply discarded at the end of their use (Armstrong 2015). Additionally, the European Union requires that at least eighty-five percent of automobile parts be recycled or reused at the end of the automotive's use (EUR-Lex 2014b). Thus, laws like these should extend with the immersion of new nanomaterials in these industries.

- **Smart and Responsive Nanocomposites**—Self-healing and shape-memory materials.
- **Enhanced Manufacturing Processes**—Scalable and cost-effective production techniques.

4.7 Conclusion

Polymer nanocomposites have emerged as a revolutionary class of materials with superior mechanical, thermal, and functional properties, making them integral to diverse applications, including aerospace, automotive, medical, and renewable energy sectors. The incorporation of nanoscale fillers enhances the structural integrity and performance of polymers, unlocking new possibilities for high-performance materials.

Despite their advantages, challenges such as nanoparticle dispersion, interfacial bonding, processing limitations, and environmental concerns must be addressed to enable large-scale commercialization. Advances in fabrication techniques, including in situ polymerization, melt blending, and solution processing, have significantly improved the dispersion and interaction of nanomaterials with polymer matrices. Additionally, polymer nanocomposites have demonstrated their potential in next-generation solar cells, offering improved charge transport, mechanical flexibility, and power conversion efficiencies.

Future research should focus on enhancing the scalability, sustainability, and safety of polymer nanocomposites while refining their functional properties. Innovations in smart nanocomposites, self-healing materials, and biodegradable polymers will further broaden their applicability. As the field progresses, continued interdisciplinary research will be essential to overcoming existing challenges and unlocking the full potential of polymer nanocomposites in advanced engineering and technological applications.

Acknowledgements We acknowledge the support of Georgia AI Manufacturing (Georgia AIM) under Award #04-79-07808 from the Economic Development Agency, US Department of Commerce. Their contribution has been instrumental in advancing our endeavors.

References

Albdiry M et al (2013) A critical review on the manufacturing processes in relation to the properties of nanoclay/polymer composites. J Compos Mater 47(9):1093–1115

Alexandre M, Dubois P (2000) Polymer-layered silicate nanocomposites: preparation, properties and uses of a new class of materials. Mater Sci Eng R Rep 28(1–2):1–63

Ali HH (2023) Efficiency enhancement of perovskite solar cells based on graphene nanocomposites as electrons and holes transport layers. Iraqi J Ind Res 10(3):47–55

Arepalli S et al (2004) Protocol for the characterization of single-wall carbon nanotube material quality. Carbon 42(8–9):1783–1791

Armstrong G (2015) An introduction to polymer nanocomposites. Eur J Phys 36(6):063001

Bogush G, Tracy M, Zukoski C Iv (1988) Preparation of monodisperse silica particles: control of size and mass fraction. J Non-Cryst Solids 104(1):95–106

Calvert P (1999) A recipe for strength. Nature 399(6733):210–211

Carrado KA, Xu L (1998) In situ synthesis of polymer−clay nanocomposites from silicate gels. Chem Mater 10(5):1440–1445

Chen G et al (2003) Preparation of polystyrene/graphite nanosheet composite. Polymer 44(6):1781–1784

Coleman JN et al (2006) Small but strong: a review of the mechanical properties of carbon nanotube–polymer composites. Carbon 44(9):1624–1652

Crosby AJ, Lee JY (2007) Polymer nanocomposites: the "nano" effect on mechanical properties. Polym Rev 47(2):217–229

Du J, Cheng HM (2012) The fabrication, properties, and uses of graphene/polymer composites. Macromol Chem Phys 213(10–11):1060–1077

Elzarka A et al (2020) Large diameter TiO $ _2 $ nanotubes enable integration of conformed hierarchical and blocking layers for enhanced dye-sensitized solar cell efficiency. arXiv:2005.01605

Fauchais P et al (1983) High pressure plasmas and their application to ceramic technology. Plasma Chem IV 59–186

Fu S, Sun Z, Huang P, Li Y, Hu N (2019) Some basic aspects of polymer nanocomposites: a critical review. Nano Mater Sci 1(1):2–30

Giannelis EP (1996) Polymer layered silicate nanocomposites. Adv Mater 8(1):29–35

Giannelis EP, Krishnamoorti R, Manias E (1999) Polymer-silicate nanocomposites: model systems for confined polymers and polymer brushes. Polym Confin Environ 107–147

Gnida P et al (2021) Impact of TiO2 nanostructures on dye-sensitized solar cells performance. Materials 14(7):1633

Herron N, Thorn DL (1998) Nanoparticles: uses and relationships to molecular cluster compounds. Adv Mater 10(15):1173–1184

Huang P et al (2016) Greatly decreased redshift and largely enhanced refractive index of monodispersed ZnO-QD/silicone nanocomposites. J Mater Chem C 4(37):8663–8669

Khan ZU, Kausar A, Ullah H (2016) A review on composite papers of graphene oxide, carbon nanotube, polymer/GO, and polymer/CNT: processing strategies, properties, and relevance. Polym Plast Technol Eng 55(6):559–581

Kingston C (2007) Challenges in the characterization of carbon nanotubes: the need for standards. In: Molecular and nanomaterial architectures group tri-national workshop on standards for nanotechnology

LeBaron PC, Wang Z, Pinnavaia TJ (1999) Polymer-layered silicate nanocomposites: an overview. Appl Clay Sci 15(1–2):11–29

Li X et al (2018) Functional MXene materials: progress of their applications. Chem An Asian J 13(19):2742–2757

Liu M et al (2014) Recent advance in research on halloysite nanotubes-polymer nanocomposite. Prog Polym Sci 39(8):1498–1525

Ma J et al (2003) A new approach to polymer/montmorillonite nanocomposites. Polymer 44(16):4619–4624
Mark J (1996) Ceramic-reinforced polymers and polymer-modified ceramics. Polym Eng Sci 36(24):2905–2920
Obrero JM et al (2024) Conformal TiO2 aerogel-like films by plasma deposition: from omniphobic antireflective coatings to perovskite solar cell photoelectrodes. ACS Appl Mater Interfaces 16(30):39745–39760
Okada A et al (1989) Nylon 6–clay hybrid. MRS Online Proceed Library (OPL) 171:45
de Oliveira AD, Beatrice CAG (2018) Polymer nanocomposites with different types of nanofiller. In: Nanocomposites-recent evolutions. IntechOpen
Oriakhi C (1998) Nano sandwiches. Chem Br 34(11):59–62
Park SK, Do Kim K, Kim HT (2002) Preparation of silica nanoparticles: determination of the optimal synthesis conditions for small and uniform particles. Colloids Surf A Physchem Eng Asp 197(1–3):7–17
Paul DR, Robeson LM (2008) Polymer nanotechnology: nanocomposites. Polymer 49(15):3187–3204
Ray SS, Okamoto M (2003) Polymer/layered silicate nanocomposites: a review from preparation to processing. Prog Polym Sci 28(11):1539–1641
Reynaud E, Gauthier C, Perez J (1999) Nanophases in polymers. Revue De Metallurgie 96(2):169–176
Sahoo NG et al (2010) Polymer nanocomposites based on functionalized carbon nanotubes. Prog Polym Sci 35(7):837–867
Saini P (2015) Fundamentals of conjugated polymer blends, copolymers and composites: synthesis, properties, and applications. John Wiley & Sons
Selema T et al (2024) Advancements in black titanium dioxide nanomaterials for solar cells: a comprehensive review. Emergent Mater 1–26
Sithole MN et al (2017) A review of semi-synthetic biopolymer complexes: modified polysaccharide nano-carriers for enhancement of oral drug bioavailability. Pharm Dev Technol 22(2):283–295
Smart L, Moore E (1995) Zeolites and related structures. Solid state chemistry. Springer, pp 238–272
Song S et al (2017) Optically tunable plasmonic two-dimensional Ag quantum dot arrays for optimal light absorption in polymer solar cells. J Phys Chem C 121(33):17569–17576
Thostenson ET, Li C, Chou T-W (2005) Nanocomposites in context. Compos Sci Technol 65(3–4):491–516
Tong F et al (2012) Flexible organic/inorganic hybrid solar cells based on conjugated polymer and ZnO nanorod array. Semicond Sci Technol 27(10):105005
Tu S et al (2018) Large dielectric constant enhancement in MXene percolative polymer composites. ACS Nano 12(4):3369–3377
Uddin F (2008) Clays, nanoclays, and montmorillonite minerals. Metall Mater Trans A 39(12):2804–2814
Vaia RA et al (1999) Polymer/layered silicate nanocomposites as high performance ablative materials. Appl Clay Sci 15(1–2):67–92
Young R, Pfender E (1985) Generation and behavior of fine particles in thermal plasmas—a review. Plasma Chem Plasma Process 5:1–37
Young RJ et al (2012) The mechanics of graphene nanocomposites: a review. Compos Sci Technol 72(12):1459–1476
Zheng W, Wong S-C (2003) Electrical conductivity and dielectric properties of PMMA/expanded graphite composites. Compos Sci Technol 63(2):225–235

Chapter 5
Polymer-Based Nanocomposites for Water Remediation: From Adsorption to Photodegradation

Harpreet Singh, Madhu Khatri, and Neha Bhardwaj

Abstract Water is a precious resource required for the sustenance of almost all life forms present on the Earth. The increase in global human population, drastic changes in global climate, fast-growing industrialization, and the presence of emerging contaminants are leading to water scarcity and pollution. Conventional water treatment technologies are inefficient in removing emerging contaminants and unable to generate potable high-quality water. Recently, the applications of polymer-based nanocomposites in water remediation are gaining attention due to their advantages such as low costs, good surface area, high stability, processability, and recyclability. These nanocomposites are usually synthesized by incorporating nanomaterials (such as metal oxide nanoparticles, carbon nanotubes, graphene, carbon dots, and nano clays) or nanoscale fillers into the polymer matrix. These nanocomposites can effectively work as adsorbents and photocatalyst materials for full-scale water treatment to remove heavy metals, organic dyes, pesticides, pharmaceutical drugs, and biological contaminants (bacteria, viruses, and fungi). The need for advances in nanotechnology for sustainable water management can be studied with the aid of these nanocomposites. This chapter focuses on the recent developments in the use of polymeric nanocomposites for water treatment and mechanisms including adsorption and photodegradation.

Keywords Adsorption · Polymer · Nanocomposites · Photodegradation · Water remediation

---

H. Singh · M. Khatri
Department of Biotechnology, University Institute of Engineering Technology (UIET), Panjab University, Chandigarh, India

N. Bhardwaj (✉)
Energy and Environment Unit, Institute of Nano Science and Technology, Mohali, Punjab, India
e-mail: nehavashisht1989@gmail.com

P. M. Visakh (ed.), *Engineering Applications of Polymer-Based Nanocomposites*, Advanced Structured Materials 237, https://doi.org/10.1007/978-3-031-93463-6_5

5.1 Introduction

Water is one of the vital sources of survival for almost all living creatures on Earth. Access to safe and potable water is one of the greatest challenges in the twenty-first century. The inadequate measures to treat urban, industrial, and agricultural wastewater result in heavily contaminated and polluted water. The rapidly increasing human population, change in global climate, fast-growing industrialization, and presence of emerging contaminants are the principal reasons responsible for the deterioration of the water quality subsequently leading to water pollution (Gogoi et al. 2018; Stefanakis and Becker 2016). Wastes from agriculture setups, dump yards, and untreated sewage are also resulting in water pollution directly or indirectly. According to a WHO report on 'Drinking Water' in 2017, nearly 50% of the world's population will face water stress by the year 2025 and 21% will have no sanitation service in less developed nations (WHO 2017). In poorly developed nations, water-borne diseases like cholera, diarrhea, and hepatitis are more prevalent (Hatami 2013). In the era of technology, nothing seems to be unachievable but various milestones are yet to be achieved in wastewater treatment. Conventional techniques such as precipitation, coagulation, flocculation, incineration, ion exchange, reverse osmosis, membrane filtration, electrochemistry, photo-electrochemistry, advanced oxidation processes (AOPs), and biological methods are used for water treatment (Khodakarami and Bagheri 2021). These processes are not efficient enough to supply pure drinking water. Further, conventional water remediation technologies are reaching their maximum limit for providing adequate water quality to meet human and environmental needs (Qu et al. 2013). The need of the time is to treat wastewater effectively and completely before releasing it into main water streams.

Recent advancements in polymer science and nanotechnology can be introduced as an effective alternative to traditional wastewater treatment processes and can develop modern water supply systems. Numerous advanced and functional materials derived from polymers have been developed and effectively utilized to enhance the processes of separation and purification (Wang et al. 2020). Polymers serve various functions such as coagulants and flocculants and aid in filtration (Moud 2022). The most frequently utilized polymers include polyethylene, polyacrylamide, poly (vinylidene fluoride), polysulfone, cellulose acetate, polyethersulfone, polycarbonates, polyacrylonitrile, and polypropylene. Nanotechnology is also rapidly growing and attracting more scientific interest due to the functional characteristics of nanomaterials (Lu et al. 2016). Nanomaterials are materials smaller than 100 nm in at least one of the dimensions, and because of their extremely small size and high surface area, nanomaterials possess large adsorption capacity and reactivity (Baig et al. 2021). Moreover, all the nanomaterials have well-defined and controllable morphologies of proper size and porosity and hence can be efficient platforms for water remediation (Li et al. 2015a).

The combination of nanomaterials and polymers in wastewater treatment is also one of the most promising approaches over other techniques (Yaqoob and Ibrahim 2019). For the remediation of contaminated water, nanocomposites provide better adsorption, selectivity, and stability than nanoparticles (NPs). Nanocomposites, in particular polymer nanocomposites, have attracted much attention among all engineered nanostructured materials. Polymer nanocomposites are materials made by incorporating nanomaterials or nanoscale fillers into the polymer matrix. These fillers are usually in the form of NPs, nanofibers, or nanolayers, which are dispersed within the polymer to enhance or modify its properties (Carpenter et al. 2015). These nanocomposites can be used as effective adsorbents, membranes, and photoreactive agents for water remediation (Adeleye et al. 2016; Otrisal et al. 2018). Different factors could affect the adsorption and photodegradation mechanisms of polymer nanocomposites to be applied in water treatment. The removal of various pollutants and contaminants from water with polymer nanocomposites is based on different phenomena such as adsorption, photocatalysis, oxidation, co-precipitation, and filtration (Chenab et al. 2020; Santhosh et al. 2016). Three important categories of nanomaterials including metal and metal oxide NPs, carbonaceous nanomaterials, and magnetic NPs are the most studied materials applied in polymer nanocomposites (Singh et al. 2020; Kumar et al. 2023; Sharma et al. 2022). Nanomaterials such as metal or metal oxides, carbon nanotubes (CNTs), and graphene possess large surface area, high reactivity, fast kinetics, and thereby exhibit specific affinity to various contaminants. This adsorption approach can be effectively used as a water purification system (Jiang et al. 2018a; Zaidi et al. 2019). On the other hand, TiO_2, CNTs, and graphene-based polymer nanocomposites are used in photocatalytic degradation, which is a more advanced technology for the breakdown of trace contaminants, organic pollutants, and microorganisms (Blanco et al. 2019).

In this chapter, we provide an overview of the preparation and properties of polymer nanocomposites and their applications in water remediation. The mechanisms related to their working from adsorption to photodegradation in water treatment are also discussed.

5.2 Polymer-Based Nanocomposites

Nanocomposites comprise NPs encapsulated in the matrix of polymer, metal, or any other matrix in multiple phases in one, two, or three dimensions (Fu et al. 2019). The properties of nanocomposites can contrast with that of the constituent materials. They exhibit distinctive physical, chemical, optical, mechanical, magnetic, and electrical characteristics and have garnered significant interest for various applications including water purification. Polymer nanocomposites are advanced materials composed of a polymer matrix embedded with nanomaterial fillers (Fig. 5.1). These fillers, which can include NPs, nanotubes, or nanoclays, are dispersed throughout the polymer (Khodakarami and Bagheri 2021). Various techniques like direct compounding, solution casting, intercalation, in situ polymerization, sol–gel synthesis, electrospinning, and electrochemical methods have been employed for creating polymer nanocomposites (Shah et al. 2022). Polymer nanocomposites can function as adsorbents, coagulants, membranes, and photocatalysts to target specific contaminants with improved efficiency and selectivity (Kundu and Karak 2022). The characterization of polymer nanocomposites involves a variety of important techniques that provide insights into their properties and behaviors. A range of methods such as thermogravimetric analysis (TGA), differential thermal analysis (DTA), differential scanning calorimetry (DSC), and thermomechanical analysis (TMA), are employed to assess the thermal properties of polymer nanocomposites (Corcione and Frigione 2012). Microscopic techniques such as scanning electron microscopy (SEM), transmission electron microscopy (TEM), and atomic force microscopy (AFM) are key techniques for visualizing the morphology and structure of polymer nanocomposites at the nanoscale. Fourier-Transform Infrared Spectroscopy (FT-IR) and Raman Spectroscopy are crucial for analyzing molecular structures and identifying functional groups within the polymer matrix and nanofillers (Ciprari et al. 2006). X-ray diffraction (XRD) is utilized to investigate the crystalline structure of nanocomposites. This technique helps in determining the phase composition, crystallite size, and overall structural integrity of the materials (Ciprari et al. 2006). The integration of polymers and nanomaterials often results in enhanced mechanical, thermal, electrical, or barrier properties. Biodegradable natural polymers present significant benefits as prospective matrix components in nanocomposites. Common natural polymers, such as polysaccharides, proteins, and lipids, possess the ability to create biocompatible nanocomposites (Zafar et al. 2016). The incorporation of a polymer matrix significantly enhances the processability of non-polymeric components within the composite, allowing for easier fabrication and manipulation (Maghimaa et al. 2024). Polymers are often selected as host materials due to their favorable processability, making them ideal candidates for creating nanocomposites. This synergy between the polymer matrix and nanomaterials underscores the importance of understanding these interactions to optimize the properties of polymer nanocomposites (Shah et al. 2022). The components of the polymer nanocomposites are discussed in the subsequent sub-sections.

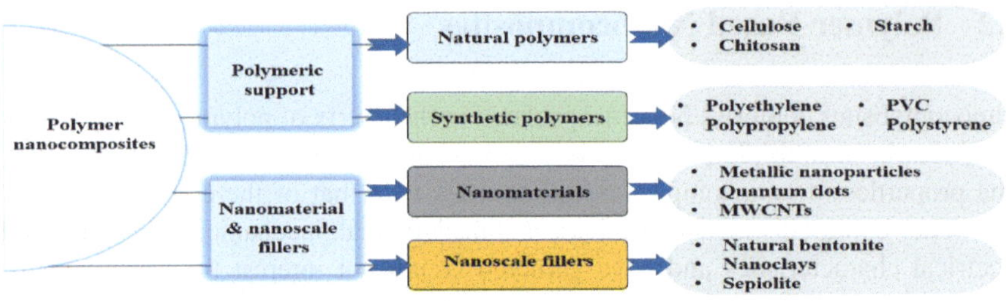

Fig. 5.1 Overview of different components of polymeric nanocomposites

5.2.1 Polymers

Polymers play a crucial role in the development of nanocomposites, serving as a versatile and adjustable matrix that imparts tailored physical and chemical properties to the resulting materials. The polymers function as the continuous phase or matrix, enabling the effective incorporation and distribution of various nanomaterials throughout the composite (Yang et al. 2024). Both natural and synthetic polymers provide unique benefits in polymeric nanocomposites. Natural polymers such as cellulose, chitosan, and starch, are advantageous due to their biodegradable and biocompatible nature as well as their origins from renewable resources. Cellulose is an abundantly available biopolymer on Earth and possesses excellent mechanical properties such as high tensile strength and flexibility (Christina et al. 2023). Its chemical structure allows for various modifications enhancing its compatibility with nanomaterials. Chitosan is derived from chitin which is found in the shells of crustaceans. It has inherent biodegradability and antimicrobial properties that make it ideal for applications in the elimination of microorganisms from water (Christina et al. 2023). Furthermore, natural polymers can be engineered to absorb pollutants or provide a matrix that supports the degradation of harmful substances. Synthetic polymers such as polyethylene, polypropylene, polyvinyl chloride, polystyrene, and epoxy resins are versatile for different applications and processing conditions (Seewoo et al. 2024). These polymers typically possess greater durability and exhibit superior thermal and mechanical stability. Synthetic biodegradable polymers like polylactic acid (PLA) and polycaprolactone (PCL) provide better alternatives toward environmental sustainability (Yang et al. 2024). Overall, the choice between natural and synthetic polymers in polymer-based nanocomposites is influenced by the need for sustainability, mechanical performance, and functionality.

5.2.2 Nanomaterials and Nanoscale Fillers

Metal and metal oxide-based nanomaterials are composed of metal and their oxides at the nanoscale. Metallic NPs are considered to be effective in decontamination due to their high reactivity, mobility, and filtration efficiency in wastewater treatment (Lv et al. 2021; Li et al. 2015b). Most studies on wastewater treatment by metallic NPs are focused on silver NPs (AgNPs), iron NPs, and Zn NPs (Neto 2019; Phenrat et al. 2020). AgNPs are regarded as strong antimicrobial agents and highly toxic to microorganisms. AgNPs provide protection against a range of microorganisms including viruses, bacteria, and fungi (Borrego et al. 2016; Prabhu and Poulose 2012). AgNPs can work as filters and membranes for the removal of microorganisms from water by incorporating cellulosic materials such as filter paper, cotton fabric, and cellulose gels. AgNPs exhibit good antibacterial properties against *Escherichia coli, Enterococcus faecalis, Pseudomonas aeruginosa, Stenotrophomonas maltophilia,* and *Acinetobacter* suspensions in water (Prabhu and Poulose 2012; Hwang et al.

2007). In addition, Zero valent iron (ZVI) is an effective material for water treatment. Its reactivity has been largely improved by the development of nanoscale ZVI (nZVI). nZVI possesses many specific positive features such as a high reactivity toward a broad range of pollutants due to their large specific areas (Fu et al. 2014; Shi et al. 2012). The nZVI can act as reducing agents due to its moderate standard reduction potential. The nZVI has successfully removed contaminants including halogenated organic compounds, nitroaromatic compounds, organic dyes, phenols, heavy metals, inorganic anions such as phosphates, and nitrates, and radio elements using the effects of adsorption, reduction, precipitation, and oxidation in the presence of dissolved oxygen (Arancibia-Miranda et al. 2016; Xiong et al. 2015; Hoag et al. 2009; Jabeen et al. 2013; Latif et al. 2020). Also, nano zero-valent zinc (nZVZ) has the potential to degrade quicker than nZVI. The majority of research on the use of nano zero-valent zinc (nZVZ) have concentrated on the dehalogenation process particularly degradation of halogenated organic compounds such as CCl_4 (Tratnyek et al. 2010). Alternatively, metal oxide NPs produced from bulk materials such as titanium dioxide (TiO_2), zinc oxide (ZnO), and iron oxides have also been used in water remediation. Metal oxide NPs have comparatively large surface area purely made of metal precursor and widely used for water purification purposes (Gold et al. 2018; Raghunath and Perumal 2017). The main mechanisms for water treatment by these materials include adsorption, chemical degradation, and photodegradation. TiO_2 is the most outstanding photocatalyst due to its strong photocatalytic properties, photostability, and cost-effectiveness (Imamura et al. 2013; Guesh et al. 2016). To generate charge separation inside the particles, TiO_2 significant bandgap energy (3.2 eV) requires UV irradiation. TiO_2 produces reactive oxygen species (ROS) in response to UV irradiation, which can destroy pollutants in a short reaction time (Umar et al. 2019; Moon et al. 2014). Because of their wide bandgap energy, they need UV stimulation, and the photocatalytic capabilities of TiO_2 NPs under visible light are relatively insignificant. Metal doping has been shown to boost the visible light absorption of TiO_2-based NPs and their photocatalytic activity under UV light (Anpo et al. 2001; Mu et al. 1989). Apart from TiO_2 NPs, ZnO NPs have also emerged as an effective choice in wastewater treatment due to their distinctive properties, such as direct bandgap in the near-UV region, proficient oxidation, and rather high photocatalytic properties (Jiang et al. 2018b; Zak et al. 2011). The ZnO NPs are environmentally beneficial since they are compatible with organisms making them ideal for wastewater treatment. Several studies have demonstrated that a combination of ZnO NPs with other semiconductors (such as CdO, CeO_2, SnO_2, and TiO_2) and reduced graphene oxide (rGO) is a viable strategy for improving the photodegradation efficiency of ZnO NPs (Gu et al. 2016; Uddin et al. 2012). Iron oxides such as magnetite (Fe_3O_4) and maghemite (γ- Fe_2O_4) have been employed as adsorbents in the removal of various heavy metals from wastewater (Lei et al. 2014; Tan et al. 2014). Magnetic nanocomposites have cores made with magnetic elements such as iron, nickel, cobalt or their oxides and different shells made of inorganic components (e.g., silica, alumina, polymers, surfactants). These coating can improve particle chemical stability, prevent oxidation, and also provide specific functionalities like selectivity for ion uptake or enhancing the water solubility of hydrophobic contaminants (Ge et al. 2018; Surendhiran et al. 2017). Carbon-based nanomaterials includes single walled carbon nanotubes (SWCNTs), multiwalled carbon nanotubes (MWCNTs), graphene, and graphene oxide (GO) have received the great attention in water remediation. CNTs are graphene folded into cylinders with an undetermined carbon-wall structure and diameter of less than 100 nm. CNTs are a part of an intriguing family due to their distinctive structures and electrical characteristics which make them appealing for adsorption processes (Abousalman-Rezvani et al. 2020; Alavi et al. 2021). Their benefits for wastewater treatment stem from their high adsorption capacity, fast kinetics, aromatic selectivity, and large surface area (Babele et al. 2021; Saifuddin et al. 2013; Vairavapandian et al. 2008). Graphene and GO possess efficient adsorption for a variety of contaminants, including dichlorobenzene, ethylbenzene, Zn^{2+}, Pb^{2+}, Cu^{2+}, and Cd^{2+}, and dyes (Sadegh et al. 2016; Ren et al. 2011; Peng et al. 2003; Lu et al. 2008; Li et al. 2003).

Nanofillers like silicates and nanoclay are also used in polymer-based nanocomposites due to their ability to significantly enhance the properties of the polymer matrix (Mattausch 2015). Silicates are naturally abundant minerals that can be incorporated into polymers to improve mechanical strength, thermal stability, and barrier properties. Nanoclays such as montmorillonite, bentonite, and hectorite possess a high cation exchange capacity, allowing for surface modifications that improve compatibility with various polymer types (Nanda et al. 2021). Through this modification, nanoclays can disperse uniformly within the polymer matrix. Silicates and nanoclays are more environmentally friendly than engineered nanomaterials thereby supporting the development of sustainable and high-performance nanocomposites (Mattausch 2015).

5.3 Properties of Polymer-Based Nanocomposites

5.3.1 Adsorption Properties

Adsorption is one of the most important steps in water remediation to remove organic and inorganic pollutants from water (Chenab et al. 2020). The traditional adsorbents suffer from certain drawbacks such as low surface area, less selectivity, and slow adsorption kinetics. Polymer nanocomposites exhibit exceptionally large specific surface area, numerous adsorption sites, and tuneable pore size (Jiang et al. 2018a). Polymer nanocomposites can be integrated into the conventional treatment process in the absorber system of reactors or wastewater treatment plants (Pan and Xing 2008). An additional step is needed to separate the polymer nanocomposites from the treated stream. Polymer nanocomposites can be used in fixed, packed, or fluidized bed reactors with pellets or porous granule beads (Umesh et al. 2024). The applications of these beds can be commercialized at large scale and pilot scales.

5.3.2 *Photocatalytic Properties*

Photocatalytic degradation is a process of absorption of the light photon to degrade a molecule or a compound (Azzam et al. 2019). The use of a light source (particularly UV light) can be used for an efficient approach to the degradation of contaminants and pollutants in water sources. The photodegradation process involves the diffusion of the contaminant (organic dye, pigment, etc.) from the solution phase to the photocatalyst and adsorption on the surface. The next step was the oxidation of the absorbed pollutant with radical species with the release of non-toxic by-products (Guesh et al. 2016). The most common radical species generated during the photocatalytic reaction is the hydroxyl (OH^-) radical when the organic contaminant is in contact with the photocatalytic material (Moradi et al. 2017; Pandey et al. 2020). The photodegradation approach is widely popular for the degradation of organic dyes and heavy metals in photocatalytic reactions. In the field of photocatalysis, polymer nanocomposites doped with TiO NPs and ZnO NPs exhibit exceptional potential for degrading organic pollutants and pathogens under light irradiation.

5.4 Applications of Polymer-Based Nanocomposites in Water Treatment

In this section, the applications of polymer-based nanocomposites in the removal of different water pollutants (such as heavy metals, organic dyes, microbes, etc.) are discussed. A significant amount of surface and groundwater is polluted with various heavy metals and organic dyes, resulting from human activities or natural processes. These heavy metal ions can be collected in consumable products. Numerous studies have investigated the removal of heavy metal ions and dyes using polymer nanocomposites. For example, thin-film nanocomposite (TFN) membranes using polyamide polymer and carboxy-functionalized TiO_2 NPs were synthesized through vapor phase interfacial polymerization (Karki and Ingole 2022). These membranes demonstrated effective ion removal achieving a rejection rate of 87% for Na^+, 86% for Cu^{2+}, 77% for Hg^{2+}, and 83% for Pb^{2+} ions. Similarly, black TiO_2 NPs were initially synthesized and integrated into the polyamide active layer of the thin-film nanofiltration membrane to enhance the membrane's performance (Mohammad Gheimasi et al. 2021). The TFN membrane exhibited superior efficiency in rejecting heavy metals, particularly demonstrating the highest rejection rates for Cr^{3+} and Pb^{2+}. In another study, polysulfone nanofiltration membranes were modified by incorporating TiO_2 NPs/dendrimers to assess their effectiveness in removing Sr^{2+}, Cu^{2+}, Pb^{2+}, and Co^{2+} as well as scientist yellow and malachite green dyes (Ranjbaran et al. 2024). Another research investigated the creation of a high-antifouling TiO_2/3-cyanopropyltriethoxysilane / Metformin-polyethersulfone (PES) membrane using different concentrations of NPs (Barahimi et al. 2020). The efficiency of the membranes was evaluated by measuring the rejection of Cu^{2+} ions, and the removal of dye from wastewater generated by a liquorice extraction plant. Also, a nanocomposite was developed by adding iron oxide-doped aluminum oxide (Fe: Al_2O_3) NPs into the polysulfone matrix (Sherugar et al. 2022). The optimized mixed matrix membrane exhibited rejection efficiencies of 99%, 98%, 62%, and 48% for Pb^{2+}, Cd^{2+}, Hg^{2+}, and F^-, respectively.

For producing a mixed matrix membrane, polysulfone was incorporated with ZnO NPs in the membrane matrix (Mondal and De 2022). It demonstrated selective adsorption of Cu^{2+} in water through surface hydroxylation, achieving a Cu^{2+} removal rate exceeding 92%. Another mixed matrix membrane was synthesized by doping aluminum oxide NPs (Zn: Al_2O_3) into polysulfone matrix (Sherugar et al. 2021). The synthesized membranes exhibited improved hydrophilicity, surface charge, and porosity, facilitating the removal of arsenic and lead with efficiencies of 87% and 98%, respectively. An investigation into the antifouling characteristics conducted at different pressures using a feed solution with Bovine Serum Albumin (BSA) demonstrated a flux recovery ratio of 98.4%. It allowed for reusability across three consecutive cycles. In another study, mixed matrix membranes composed of poly (vinylidene fluoride) (PVDF) and alpha-zirconium phosphate (PVDF-α-ZrP) were created using the phase inversion technique (Abdulkarem et al. 2021). The PVDF-α-ZrP membrane that included 1.00 wt % α-ZrP demonstrated the ability to eliminate 42.8% (Cd^{2+}), 93.1% (Cu^{2+}), 44.4% (Ni^{2+}), 91.2% (Pb^{2+}), and 44.2% (Zn^{2+}) from aqueous solutions. Another study reported ion exchange membranes made from polyvinylidene fluoride and SnO_2 NPs (Ibrahim et al. 2020). The membrane containing 0.25 wt% PVDF/SnO_2 exhibited the greatest tensile strength of 21.3 MPa, in contrast to the pristine PVDF membrane, which showed a strength of only 14.3 MPa. Furthermore, membranes achieved maximum removal rates for Pb^{2+}, Cu^{2+}, Zn^{2+}, Cd^{2+}, and Ni^{2+} reaching approximately 93.9%, 92.8%, 82.3%, 70.7%, and 63.9%, respectively. Also, hydrous manganese oxide NPs were integrated into polyvinyl chloride nanofiber matrices via an electrospinning technique (Hezarjaribi et al. 2020). The resulting electrospun nanomaterials were able to remove Cu^{2+} and Ni^{2+} ions achieving removal rates of 90% and 80%, respectively. One study focused on the fabrication of polyvinylidene fluoride (PVDF)-based nanocomposite modified with ZnO and poly-dopamine (PDA) (Üçel and Demirel 2022). The ZnO and ZnO/PDA powders were incorporated into the membrane matrix in specific proportions using the blending method, while PDA was applied by coating. The PVDF/ZnO/PDA membrane demonstrated the highest separation efficiency for the removal of the Pb^{2+}-chitosan complex, achieving 88.5% removal rates. Another study developed polythiocyanuric acid modified with MoS_2 nanosheets (Sreeramareddygari et al. 2021). This modification allowed the materials to adhere strongly to polyethersulfone (PES) membranes without needing to be previously exfoliated. The diffusion of charges on the binding sites of the membrane contributed to the effective removal of Hg^{2+}, Cr^{6+}, Pb^{2+}, and congo red dye. The elimination of congo red was due to electrostatic repulsion, whereas the removal of the heavy metals occurred through adsorption onto the membrane.

Natural polymer-based nanocomposites have gained recognition as effective materials due to their environment-friendly nature. In this study, the starch-ZnO nanocomposite comprised a suspension of cross-linked starch and a gel of ZnO

(Naushad et al. 2021). This was utilized as a multifunctional adsorbent to extract toxic Pb^{2+} ions from the aqueous environment. The resulting starch-ZnO nanocomposite exhibited rapid adsorption equilibrium for Pb^{2+} within 120 min. In another study, a nanocomposite membrane was developed by combining Fe_3O_4@SiO_2–chitosan (Kamari and Shahbazi 2021). This composite was created using a core–shell structure of Fe_3O_4 magnetic NPs and amorphous SiO_2 derived from wheat straw, which was then modified with chitosan groups. The heavy metal ion removal efficiency of the membranes was 98%, 98%, and 98% for Pb^{2+}, Cu^{2+}, and Cd^{2+}, respectively. Additionally, the membranes exhibited over 98.2% retention rates for various dyes including cationic and anionic types. Similarly, Fe_3O_4@SiO_2–NH_2 nanocomposite was incorporated into polyethersulfone membranes at varying concentrations using the phase inversion technique (Kamari and Shahbazi 2020) The effectiveness of the membranes was assessed by evaluating the pure water flux, rejection of salts, removal of Cd^{2+} ions, retention of methyl red dye, and antifouling properties. The nanocomposite at a concentration of 0.5 wt % demonstrated the best efficiency for eliminating Cd^{2+} ions (93%) and methyl red dye (97%).

Carbon-based nanomaterials show significant potential for creating advanced adsorptive nanocomposite membranes that are highly effective in the remediation of water pollutants. For example, carboxylic acid functionalized MWCNTs (MWCNTs-COOH) were used as an additive to introduce the COOH functional group into the chitosan matrix (Alawady et al. 2020). Two nanocomposites namely CNTs-COOH/chitosan and CNTs/chitosan were used as a selective layer placed on top of polysulfone to create CNTs-COOH/chitosan/PS and CNTs/chitosan/PS membranes. Both CNTs-COOH/chitosan/PS and CNTs/chitosan/PS membranes effectively remove Cu^{2+}, Ni^{2+}, Pb^{2+}, Cd^{2+}, and Co^{2+} ions from water solutions. At a pH of 3, the CNTs-COOH/chitosan/PS and CNTs/chitosan/PS membranes exhibited significant rejection rates ranging from 71% to 92.2% for the metal ions. In a pH range of 6 to 8, the rejection of Co^{2+} reached up to 92.2% and 89.6% for the CNTs/chitosan/PS and CNTs-COOH/chitosan/PS membranes, respectively. Similarly, ultrafiltration flat-sheet membranes were created using polyphenylsulfone, MWCNTs, and polyvinylpyrrolidone (Nayak et al. 2020). The membrane demonstrated removal efficiencies of 98%, 76%, and 72% for Pb^{2+}, Hg^{2+}, and Cd^{2+} ions, respectively. In another study, forward osmosis membranes were created with a support layer that included various concentrations (0–8 wt %) of polyethylene glycol 400, polysulfone, and 1-methyl, 2-pyrrolidone alongside an active layer which was formed by integrating different weight ratios (0–0.012 wt%) of graphene oxide (GO), 1, 3-phenylenediamine, and 1, 3, 5-benzene trichloride into the polyamide layer (Saeedi-Jurkuyeh et al. 2020). The TFN FO membranes exhibited rejection rates for Pb, Cd, and Cr of 99.9%, 99.7%, and 98.3%, respectively. Another research investigated the synthesis of a new nanohybrid composed of GO and manganese oxide and its integration into sulfonated polyethersulfone (Ibrahim et al. 2022). The rejection rates for heavy metal ions were 81.1%, 64.0%, and 67.4% for Cu^{2+}, Zn^{2+}, and Ni^{2+} ions, respectively. A study synthesized polyvinylchloride (PVC) nanocomposite membranes using aspartic acid (Asp) functionalized GO nanolayers, capable of adsorption (Namdar et al. 2021). The nanolayers exhibited the greatest performance regarding permeation, chromium removal, humic acid rejection, and antifouling properties over five filtration and regeneration cycles. The simultaneous presence of both chromium and humic acid positively influenced the membrane filtration and regeneration due to the formation of the chromium–humic acid complex through electrostatic interactions. Furthermore, GO was integrated into the polyamide layer, supported by a polysulfone membrane containing cyclodextrin (Badmus et al. 2021). Flux and separation experiments were performed to assess the permeability and separation capabilities of the developed composite membrane for eliminating salts, heavy metals, and organic pollutants. The findings indicate that the membrane achieved a peak flux of 100 $L\ m^{-2}\ h^{-1}$ within 15 min, which subsequently decreased and then stabilized at 40 $L\ m^{-2}\ h^{-1}$ after 4 h. Nearly complete rejection was observed in the removal of 1000 mg/L, 100 mg/L, and 250 mg/L of salts, heavy metals (Pb and Ni), and organic pollutants over a duration of 5 h. Later on, surface modification of GO nanoplates was conducted and utilized for the development of nanofiltration membranes based on polyetherimide (Bandehali et al. 2020). The rejection rates for Na_2SO_4, $Pb(NO_3)_2$, $CrSO_4$, and $Cu(NO_3)_2$ were significantly improved when compared to those of pure membranes. In another report, two different methods were employed to introduce functionalized GO onto the surface of nanofiltration membranes in order to enhance their performance in heavy metal removal (Lari et al. 2022). The first method involved coating the surface of the membrane with cross-linked GO, while the second method integrated GO into the monomer solution during the interfacial polymerization process. Both modified membranes demonstrated higher water flux compared to the control membrane without GO incorporation, achieving a flux of 40 L/m^2 h. It achieved a higher cobalt removal efficiency of 97%. Recently, thin-film nanocomposite membranes were developed using the interfacial polymerization technique between trimesoyl chloride and m-phenylenediamine monomers on the upper surface of a polyether sulfone membrane (Fouladi et al. 2023). These membranes were enhanced with GO and aluminum fumarate (AlFu) metal–organic framework (MOF) to facilitate the removal of Cr^{2+}, Cu^{2+}, and Ni^{2+} ions. In another study, a polyamide thin-film composite membrane was developed through the method of interfacial polymerization utilizing polysulfone as the substrate material (Masenye et al. 2022). This process involved the use of hyperbranched polyethyleneimine-functionalized MWCNTs (HPEI/MWCNTs) in conjunction with trimesoyl chloride. The resulting nanocomposite membrane was assessed for its efficacy in adsorbing Pb^{2+} ions from solution with the removal of more than 95% of Pb^{2+} ions within 10 min. For dye desalination applications, a study utilized the chelation properties of polyethyleneimine (PEI) and alginate with metal ions to create loose nanofiltration membranes (Metecan et al. 2021). The positively charged polyamide-imide (PAI)/PEI support served as the foundation for this membrane. The alginate that was applied to the support enabled the formation of permanent assemblies after being cross-linked with Fe^{3+} and Zn^{2+} metal ions. The PAI/PEI-Alg-Fe^{3+} membrane demonstrated the highest permeability, excellent resistance to fouling when exposed to synthetic textile wastewater, and sustained long-term stability in both static and dynamic environments. Additionally, the PAI/PEI-Alg-Fe^{3+} membrane showed good efficiency for desalinating methyl green and other

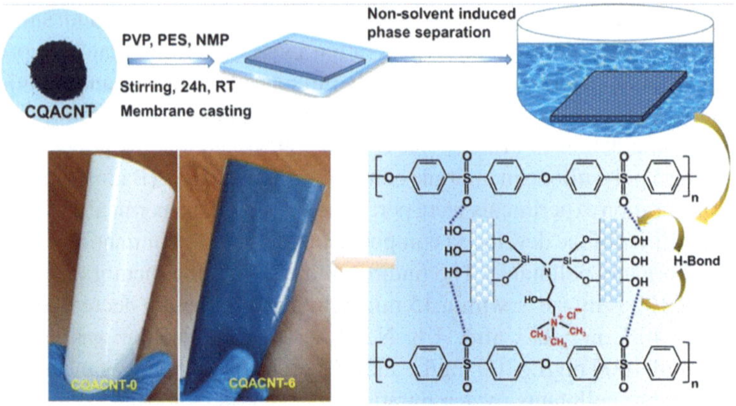

Fig. 5.2 Membrane fabrication process, chemical structure of the membrane, possible interactions between PES and CQACNT (cross-linked multiwalled carbon nanotube nanomaterial), and optical images of the membranes (Pandey et al. 2022)

similarly sized dyes. In a report, bis[3-(trimethoxysilyl)propyl]amine (BTMSPA) cross-linked MWCNT nanomaterial (CQACNT) was created through the silanization of MWCNT-OH, followed by the attachment of positively charged quaternary ammonium groups (Pandey et al. 2022). The composite membranes were formed by integrating CQACNT into the polyether sulfone polymer matrix (Fig. 5.2). The composite membrane demonstrated impressive heavy metal removal efficiencies (Pb: 89.53%; Ni: 90.42%; Cu: 91.43%; Zn: 91.86%).

Polymer-based nanocomposites have also worked as efficient photocatalysts with great importance with activity against several potential pollutants including microbes. For instance, a nanocomposite based on melamine–formaldehyde polymer and decorated with silver phosphate (Ag_3PO_4) NPs was synthesized (Garg et al. 2023). This synthesized nanocomposite was utilized for the photocatalytic degradation of bisphenol A (BPA) when exposed to sunlight. Additionally, the antibacterial activity of the nanocomposite was assessed based on concentration against two bacterial strains (*Escherichia coli* and *Bacillus subtilis*) demonstrating significant antibacterial activity. In another study, nanocomposites were created by combining aqueous $Mg(OH)_2$ with poly(vinyl alcohol) and poly(vinylpyrrolidone) both serving as the polymer matrix (Dharamalingam et al. 2023). Under UV–visible light irradiation, the nanocomposite demonstrated the highest photocatalytic efficiency, successfully degrading 93% of the methylene blue solution (20 ppm) within 2 h. $Mg(OH)_2$-poly(vinyl alcohol) maintained consistent efficiency over four reuse cycles. The antibacterial properties were evaluated using the *E. coli* MTCC 443 strain. Among all samples, $Mg(OH)_2$-poly(vinyl alcohol) showed the most significant antibacterial effect producing a zone of inhibition measuring 18 mm with a minimum inhibitory concentration of 4 mg/mL. Another photocatalyst was developed using a combination of chitosan and high-purity ZnO NPs (Aouadi et al. 2024). The size of the ZnO NPs was recorded at 12 nm whereas the chitosan–ZnO NPs measured 21 nm with bandgap energies of 3.98 and 3.48 for these materials, respectively. ZnO NPs, chitosan–ZnO NPs, and chitosan–ZnO/PVP exhibited significant antibacterial properties, particularly against Gram-positive bacteria. At a concentration of 6×10^{-5} M, chitosan achieved a 90.5% removal of methylene blue dye, while ZnO NPs removed 97.4% of methylene blue, but chitosan–ZnO/PVP achieved a complete removal of 100%. For toluidine blue (TB) at a concentration of 4×10^{-3} M, the corresponding removal efficiencies were 96.8%, 96.8%, and 100%, respectively. Similarly, a straightforward synthesis approach was employed to fabricate gallotannin-loaded chitosan/ZnO nanocomposite (Bharathi et al. 2023). The resulting nanocomposite was tested for its ability to degrade methylene blue in an aqueous solution. After 30 min of exposure to irradiation, the degradation efficiency of the nanocomposite was determined to be 96.64%. Additionally, the prepared nanocomposite exhibited potential and concentration-dependent antibacterial properties against *S. aureus*. Furthermore, a study developed photocatalytic polymeric nanocomposites using a simple approach involving sonication and solution casting (Cantarella et al. 2024). In this process, poly(methyl methacrylate) served as the supporting matrix, while ZnO NPs act as the photoactive component, and MoS_2 nanoflakes functioned as a co-catalyst for forming the heterojunction. The photocatalytic efficiencies of the resulting materials were tested by examining the photodegradation of rhodamine B (RhB) and sodium dodecyl sulfate (SDS) under UV light irradiation. Additionally, the antibacterial properties of the nanocomposites were investigated using *E. co*li as the model organism. After an hour of exposure to the nanocomposites, the survival rate of the bacteria decreased by approximately 35% with the ZnO-poly(methyl methacrylate) composite. Table 5.1 enlists some of the important studies in the field of polymer nanocomposites for water treatment.

5.5 Conclusion

The area of polymer nanocomposites is currently pacing up at a fast scale as the polymers and nanomaterials show great potential in wastewater treatment processes. Polymer nanocomposites possess adsorptive and photocatalytic properties that make them one of the best materials for the removal of contaminants such as heavy metal ions, organic dyes, pesticides, microbes, and other chemical contaminants from water. The introduction of polymer nanocomposites into wastewater treatment processes has gained interest due to their flexibility, high surface area, and the ability to selectively target specific pollutants. Metal and metal oxide NPs like silver, titanium dioxide, and iron oxide are embedded in polymer matrices due to their antimicrobial, photocatalytic, and adsorptive capabilities. The adsorption process depends upon the number of active sites present on the adsorbent and the mechanism of interaction between the adsorbent and adsorbate. The photocatalytic approach is based on the kinetics of the photocatalytic reaction, hole–electron recombination, production of radical species, and light-driven ability to photocatalyze the process. The time of hole–electron recombination pairs of photocatalysts can restrict the concentration of the radical

Table 5.1 Compilation of some studies showing the applications of polymer nanocomposites in the remediation of different contaminants from water

S. no	Polymeric nanocomposite		Target contaminant	Method of remediation	Reference
	Polymer substrate	Nanomaterial/nanoscale filler			
•	Polyamide	TiO_2 NPs	Na^+, Cu^{2+}, Hg^{2+}, Pb^{2+}	Membrane filtration	Karki and Ingole (2022)
•	Polyamide	TiO_2 NPs	Cr^{3+} and Pb^{2+}	Membrane filtration	Mohammad Gheimasi et al. (2021)
•	Polysulfone	TiO_2 NPs	Sr^{2+}, Cu^{2+}, Pb^{2+}, Co^{2+}, scientist yellow and malachite green	Membrane filtration	Ranjbaran et al. (2024)
•	Polyethersulfone	TiO_2	Cu^{2+}	Membrane filtration	Barahimi et al. (2020)
•	Polysulfone	$Fe{:}Al_2O_3$	Pb^{2+}, Cd^{2+}, Hg^{2+}, and F^-	Membrane filtration	Sherugar et al. (2022)
•	Polysulfone	ZnO NPs	Cu^{2+}	Membrane filtration	Mondal and De (2022)
•	Polysulfone	$Zn{:}Al_2O_3$	As^{3+}, Pb^{2+}	Membrane filtration	Sherugar et al. (2021)
•	Polyvinylidene fluoride	α-ZrP	Cd^{2+}, Cu^{2+}, Ni^{2+}, Pb^{2+}, Zn^{2+}	Membrane filtration	Abdulkarem et al. (2021)
•	Polyvinylidene fluoride	SnO_2 NPs	Pb^{2+}, Cu^{2+}, Zn^{2+}, Cd^{2+}, Ni^{2+}	Membrane filtration	Ibrahim et al. (2020)
•	Polyvinyl chloride	MnO_2 NPs	Cu^{2+}, Ni^{2+}	Membrane filtration	Hezarjaribi et al. (2020)
•	Polyvinylidene fluoride	ZnO/PDA	Pb^{2+}	Membrane filtration	Üçel and Demirel (2022)
•	Polyethersulfone	MoS_2 nanosheets	Hg^{2+}, Cr^{6+}, Pb^{2+}, congo red	Membrane filtration	Sreeramareddygari et al. (2021)
•	Starch	ZnO	Pb^{2+}	Adsorption	Naushad et al. (2021)

(continued)

Table 5.1 (continued)

S. no	Polymeric nanocomposite		Target contaminant	Method of remediation	Reference
	Polymer substrate	Nanomaterial/nanoscale filler			
•	Chitosan	$Fe_3O_4@SiO_2$	Pb^{2+}, Cu^{2+}, Cd^{2+}, cationic and anionic dyes	Membrane filtration	Kamari and Shahbazi (2021)
•	Polyethersulfone	$Fe_3O_4@SiO_2\text{-}NH_2$	Cd^{2+}, methyl red	Adsorption	Kamari and Shahbazi (2020)
•	Chitosan/polysulfone	MWCNTs	Cu^{2+}, Ni^{2+}, Pb^{2+}, Cd^{2+}, Co^{2+}	Membrane filtration	Alawady et al. (2020)
•	Polyvinylpyrrolidone/polyphenylsulfone	MWCNTs	Pb^{2+}, Hg^{2+}, Cd^{2+}	Membrane filtration	Nayak et al. (2020)
•	Polyamide	GO	Pb^{2+}, Cd^{2+}, Cr^{2+}	Membrane filtration	Saeedi-Jurkuyeh et al. (2020)
•	Polyethersulfone	GO	Cu^{2+}, Zn^{2+}, Ni^{2+}	Membrane filtration	Ibrahim et al. (2022)
•	Polyvinylchloride	GO	Cr^{2+}, humic acid	Membrane filtration	Namdar et al. (2021)
•	Polyamide/polysulfone	GO	Pb^{2+}, Ni^{2+}	Membrane filtration	Badmus et al. (2021)
•	Polyetherimide	GO	Na_2SO_4, $Pb(NO_3)_2$, $CrSO_4$, $Cu(NO_3)_2$	Membrane filtration	Bandehali et al. (2020)
•	Polysulfone	GO	Cobalt	Membrane filtration	Lari et al. (2022)
•	Polyethersulfone	GO/Al-Fu MOF	Cr^{2+}, Cu^{2+}, Ni^{2+}	Membrane filtration	Fouladi et al. (2023)
•	Polyamide	MWCNTs	Pb^{2+}	Membrane filtration	Masenye et al. (2022)
•	Polyamide/alginate	PEI	Fe^{3+}, Zn^{2+}, methyl green	Membrane filtration	Metecan et al. (2021)

(continued)

Table 5.1 (continued)

S. no	Polymeric nanocomposite		Target contaminant	Method of remediation	Reference
	Polymer substrate	Nanomaterial/nanoscale filler			
•	Polyether sulfone	MWCNTs	Pb^{2+}, Ni^{2+}, Cu^{2+}, Zn^{2+}	Membrane filtration	Pandey et al. (2022)
•	Melamine–formaldehyde	Ag_3PO_4 NPs	BPA, *E. coli, Bacillus subtilis*	Photocatalytic degradation	Garg et al. (2023)
•	Poly(vinyl alcohol) and poly(vinylpyrrolidone)	$Mg(OH)_2$	Methylene blue, *E. coli* MTCC 443 strain	Photocatalytic degradation	Dharamalingam et al. (2023)
•	Chitosan	ZnO NPs	Methylene blue, toluidine blue	Photocatalytic degradation	Aouadi et al. (2024)
•	Chitosan	ZnO NPs	Methylene blue, *S aureus*	Photocatalytic degradation	Bharathi et al. (2023)
•	Poly(methyl methacrylate) s	ZnO NPs	Rhodamine B, sodium dodecyl sulfate, *E. coli*	Photocatalytic degradation	Cantarella et al. (2024)

spices. Many of the polymer-based nanocomposites in this field are at the laboratory scale research. Further developments in enhancing their recyclability, biodegradability, and pollutant specificity are expected to push polymer nanocomposites to the forefront of wastewater treatment technologies in the coming years.

Acknowledgements This work is supported by DST INSPIRE faculty grant (Reg. no. IFA18-LSPA 127) awarded to Dr. Neha Bhardwaj at the Institute of Nano Science and Technology (INST), Mohali, Punjab.

Declaration of Competing Interest The authors declare that they have NO known competing interests to declare.

References

Abdulkarem E, Ibrahim Y, Kumar M, Arafat HA, Naddeo V, Banat F, Hasan SW (2021) Polyvinylidene fluoride (PVDF)-α-zirconium phosphate (α-ZrP) nanoparticles based mixed matrix membranes for removal of heavy metal ions. Chemosphere 267:128896

Abousalman-Rezvani Z, Eskandari P, Roghani-Mamaqani H, Salami-Kalajahi M (2020) Functionalization of carbon nanotubes by combination of controlled radical polymerization and "grafting to" method. Adv Coll Interface Sci 278:102126

Adeleye AS, Conway JR, Garner K, Huang Y, Su Y, Keller AA (2016) Engineered nanomaterials for water treatment and remediation: costs, benefits, and applicability. Chem Eng J 286:640–662

Alavi M, Jabari E, Jabbari E (2021) Functionalized carbon-based nanomaterials and quantum dots with antibacterial activity: a review. Expert Rev Anti Infect Ther 19(1):35–44

Alawady AR, Alshahrani AA, Aouak TA, Alandis NM (2020) Polysulfone membranes with CNTs/Chitosan biopolymer nanocomposite as selective layer for remarkable heavy metal ions rejection capacity. Chem Eng J 388:124267

Anpo M, Kishiguchi S, Ichihashi Y, Takeuchi M, Yamashita H, Ikeue K, Morin B, Davidson A, Che M (2001) The design and development of second-generation titanium oxide photocatalysts able to operate under visible light irradiation by applying a metal ion-implantation method. Res Chem Intermed 27

Aouadi A, Hamada Saud D, Rebiai A, Achouri A, Benabdesselam S, Mohamed Abd El-Mordy F, Pohl P, Ahmad SF, Attia SM, Abulkhair HS (2024) Introducing the antibacterial and photocatalytic degradation potentials of biosynthesized chitosan, chitosan–ZnO, and chitosan–ZnO/PVP nanoparticles. Sci Rep 14(1):14753

Arancibia-Miranda N, Baltazar SE, García A, Muñoz-Lira D, Sepúlveda P, Rubio MA, Altbir D (2016) Nanoscale zero valent supported by zeolite and montmorillonite: template effect of the removal of lead ion from an aqueous solution. J Hazard Mater 301:371–380

Azzam EM, Fathy NA, El-Khouly SM, Sami RM (2019) Enhancement the photocatalytic degradation of methylene blue dye using fabricated CNTs/TiO2/AgNPs/Surfactant nanocomposites. J Water Process Eng 28:311–321

Babele PK, Verma MK, Bhatia RK (2021) Carbon nanotubes: a review on risks assessment, mechanism of toxicity and future directives to prevent health implication. Biocell 45(2):267

Badmus SO, Oyehan TA, Saleh TA (2021) Enhanced efficiency of polyamide membranes by incorporating cyclodextrin-graphene oxide for water purification. J Mol Liq 340:116991

Baig N, Kammakakam I, Falath W (2021) Nanomaterials: a review of synthesis methods, properties, recent progress, and challenges. Mater Adv 2(6):1821–1871

Bandehali S, Moghadassi A, Parvizian F, Zhang Y, Hosseini SM, Shen J (2020) New mixed matrix PEI nanofiltration membrane decorated by glycidyl-POSS functionalized graphene oxide nanoplates with enhanced separation and antifouling behaviour: Heavy metal ions removal. Sep Purif Technol 242:116745

Barahimi V, Taheri RA, Mazaheri A, Moghimi H (2020) Fabrication of a novel antifouling TiO 2/CPTES/metformin-PES nanocomposite membrane for removal of various organic pollutants and heavy metal ions from wastewater. Chem Pap 74:3545–3556

Bharathi D, Dhanasekaran S, Varshini R, Bhuvaneswari S, Periyasami G, Pandiaraj S, Lee J, Ranjithkumar R (2023) Preparation of gallotannin loaded chitosan/zinc oxide nanocomposite for photocatalytic degradation of organic dye and antibacterial applications. Int J Biol Macromol 243:125052

Blanco M, Monteserín C, Angulo A, Pérez-Márquez A, Maudes J, Murillo N, Aranzabe E, Ruiz-Rubio L, Vilas JL (2019) TiO2-doped electrospun nanofibrous membrane for photocatalytic water treatment. Polymers 11(5):747

Borrego B, Lorenzo G, Mota-Morales JD, Almanza-Reyes H, Mateos F, López-Gil E, de la Losa N, Burmistrov VA, Pestryakov AN, Brun A (2016) Potential application of silver nanoparticles to control the infectivity of Rift Valley fever virus in vitro and in vivo, nanomedicine: nanotechnology. Biol Med 12(5):1185–1192

Cantarella M, Spanò V, Zimbone M, Giuffrida F, Lufrano E, Strano V, Franzò G, Sfuncia G, Nicotra G, Alberti A (2024) ZnO–MoS2-PMMA polymeric nanocomposites: a harmless material for water treatment. Mater Today Chem 36:101912

Carpenter AW, de Lannoy C-F, Wiesner MR (2015) Cellulose nanomaterials in water treatment technologies. Environ Sci Technol 49(9):5277–5287

Chenab KK, Sohrabi B, Jafari A, Ramakrishna S (2020) Water treatment: functional nanomaterials and applications from adsorption to photodegradation. Mater Today Chem 16:100262

Christina K, Subbiah K, Arulraj P, Krishnan SK, Sathishkumar P (2023) A sustainable and eco-friendly approach for environmental and energy management using biopolymers chitosan, lignin and cellulose—a review. Int J Biol Macromol 128550

Ciprari D, Jacob K, Tannenbaum R (2006) Characterization of polymer nanocomposite interphase and its impact on mechanical properties. Macromolecules 39(19):6565–6573

Corcione CE, Frigione M (2012) Characterization of nanocomposites by thermal analysis. Materials 5(12):2960–2980

Dharamalingam K, Ramasundaram S, Ponnusamy VK, Bhuvaneswari K, Ramalingam G, Balasankar A, Jeyaram S, Pazhanivel T, Florence SS, Thangavel E (2023) Facile synthesis of polymer-based magnesium hydroxide nanocomposites for photocatalytic degradation for methylene blue dye and antibacterial application. Biomass Convers Biorefinery 13(15):13539–13552

Fouladi M, Kavousi Heidari M, Tavakoli O (2023) Performance comparison of thin-film nanocomposite polyamide nanofiltration membranes for heavy metal/salt wastewater treatment. J Nanoparticle Res 25(4):77

Fu F, Dionysiou DD, Liu H (2014) The use of zero-valent iron for groundwater remediation and wastewater treatment: a review. J Hazard Mater 267:194–205

Fu S, Sun Z, Huang P, Li Y, Hu N (2019) Some basic aspects of polymer nanocomposites: a critical review. Nano Mater Sci 1(1):2–30

Garg AK, Aggarwal R, Kulshreshtha NM, Dalal C, Gupta K, Sonkar SK (2023) Ag3PO4Nanoparticles-decorated melamine-formaldehyde polymer nanocomposite as a catalyst for the photodegradation of bisphenol a and its antibacterial activity. ACS Appl Nano Mater 6(22):20909–20918

Ge L, Wang W, Peng Z, Tan F, Wang X, Chen J, Qiao X (2018) Facile fabrication of Fe@ MgO magnetic nanocomposites for efficient removal of heavy metal ion and dye from water. Powder Technol 326:393–401

Gogoi A, Mazumder P, Tyagi VK, Chaminda GT, An AK, Kumar M (2018) Occurrence and fate of emerging contaminants in water environment: a review. Groundw Sustain Dev 6:169–180

Gold K, Slay B, Knackstedt M, Gaharwar AK (2018) Antimicrobial activity of metal and metal-oxide based nanoparticles. Adv Therapeutics 1(3):1700033

Gu X, Li C, Yuan S, Ma M, Qiang Y, Zhu J (2016) ZnO based heterojunctions and their application in environmental photocatalysis. Nanotechnology 27(40):402001

Guesh K, Mayoral A, Marquez-Alvarez C, Chebude Y, Diaz I (2016) Enhanced photocatalytic activity of TiO_2 supported on zeolites tested in real wastewaters from the textile industry of Ethiopia. Microporous Mesoporous Mater 225:88–97

Hatami H (2013) Importance of water and water-borne diseases: on the occasion of the world water day (march 22, 2013). Int J Prev Med 4(3):243

Hezarjaribi M, Bakeri G, Sillanpää M, Chaichi M, Akbari S (2020) Novel adsorptive membrane through embedding thiol-functionalized hydrous manganese oxide into PVC electrospun nanofiber for dynamic removal of Cu (II) and Ni (II) ions from aqueous solution. J Water Process Eng 37:101401

Hoag GE, Collins JB, Holcomb JL, Hoag JR, Nadagouda MN, Varma RS (2009) Degradation of bromothymol blue by 'greener'nano-scale zero-valent iron synthesized using tea polyphenols. J Mater Chem 19(45):8671–8677

Hwang MG, Katayama H, Ohgaki S (2007) Inactivation of Legionella pneumophila and Pseudomonas aeruginosa: evaluation of the bactericidal ability of silver cations. Water Res 41(18):4097–4104

Ibrahim Y, Naddeo V, Banat F, Hasan SW (2020) Preparation of novel polyvinylidene fluoride (PVDF)-Tin (IV) oxide (SnO2) ion exchange mixed matrix membranes for the removal of heavy metals from aqueous solutions. Sep Purif Technol 250:117250

Ibrahim Y, Wadi VS, Ouda M, Naddeo V, Banat F, Hasan SW (2022) Highly selective heavy metal ions membranes combining sulfonated polyethersulfone and self-assembled manganese oxide nanosheets on positively functionalized graphene oxide nanosheets. Chem Eng J 428:131267

Imamura K, Yoshikawa T, Hashimoto K, Kominami H (2013) Stoichiometric production of aminobenzenes and ketones by photocatalytic reduction of nitrobenzenes in secondary alcoholic suspension of titanium (IV) oxide under metal-free conditions. Appl Catal B 134:193–197

Jabeen H, Kemp KC, Chandra V (2013) Synthesis of nano zerovalent iron nanoparticles–graphene composite for the treatment of lead contaminated water. J Environ Manage 130:429–435

Jiang N, Shang R, Heijman SG, Rietveld LC (2018a) High-silica zeolites for adsorption of organic micro-pollutants in water treatment: a review. Water Res 144:145–161

Jiang J, Pi J, Cai J (2018) The advancing of zinc oxide nanoparticles for biomedical applications. Bioinorg Chem Appl

Kamari S, Shahbazi A (2020) Biocompatible Fe3O4@ SiO2-NH2 nanocomposite as a green nanofiller embedded in PES–nanofiltration membrane matrix for salts, heavy metal ion and dye removal: long–term operation and reusability tests. Chemosphere 243:125282

Kamari S, Shahbazi A (2021) High–performance nanofiltration membrane blended by Fe3O4@ SiO2–CS bionanocomposite for efficient simultaneous rejection of salts/heavy metals ions/dyes with high permeability, retention increase and fouling decline. Chem Eng J 417:127930

Karki S, Ingole PG (2022) Development of polymer-based new high performance thin-film nanocomposite nanofiltration membranes by vapor phase interfacial polymerization for the removal of heavy metal ions. Chem Eng J 446:137303

Khodakarami M, Bagheri M (2021) Recent advances in synthesis and application of polymer nanocomposites for water and wastewater treatment. J Clean Prod 296:126404

Kumar D, Singh H, Jain A, Sharma V, Bhardwaj N, Puri S, Khatri M (2023) Response surface methodology for removal of copper (II) ions from aqueous solutions by DMSA@ SiO2@ Fe3O4 nanocomposite. Chem Pap 77(4):1907–1920

Kundu S, Karak N (2022) Polymeric photocatalytic membrane: an emerging solution for environmental remediation. Chem Eng J 438:135575

Lari S, Parsa S, Akbari S, Emadzadeh D, Lau W (2022) Fabrication and evaluation of nanofiltration membrane coated with amino-functionalized graphene oxide for highly efficient heavy metal removal. Int J Environ Sci Technol 19(6):4615–4626

Latif A, Sheng D, Sun K, Si Y, Azeem M, Abbas A, Bilal M (2020) Remediation of heavy metals polluted environment using Fe-based nanoparticles: mechanisms, influencing factors, and environmental implications. Environ Pollut 114728
Lei Y, Chen F, Luo Y, Zhang L (2014) Three-dimensional magnetic graphene oxide foam/Fe 3 O 4 nanocomposite as an efficient absorbent for Cr (VI) removal. J Mater Sci 49(12):4236–4245
Li Y-H, Ding J, Luan Z, Di Z, Zhu Y, Xu C, Wu D, Wei B (2003) Competitive adsorption of Pb2+, Cu2+ and Cd2+ ions from aqueous solutions by multiwalled carbon nanotubes. Carbon 41(14):2787–2792
Li R, Zhang L, Wang P (2015a) Rational design of nanomaterials for water treatment. Nanoscale 7(41):17167–17194
Li Z-J, Wang L, Yuan L-Y, Xiao C-L, Mei L, Zheng L-R, Zhang J, Yang J-H, Zhao Y-L, Zhu Z-T (2015b) Efficient removal of uranium from aqueous solution by zero-valent iron nanoparticle and its graphene composite. J Hazard Mater 290:26–33
Lu C, Su F, Hu S (2008) Surface modification of carbon nanotubes for enhancing BTEX adsorption from aqueous solutions. Appl Surf Sci 254(21):7035–7041
Lu H, Wang J, Stoller M, Wang T, Bao Y, Hao H (2016) An overview of nanomaterials for water and wastewater treatment. Adv Mater Sci Eng
Lv H, Niu H, Zhao X, Cai Y, Wu F (2021) Carbon zero-valent iron materials possessing high-content fine Fe0 nanoparticles with enhanced microelectrolysis-Fenton-like catalytic performance for water purification. Appl Catal B 286:119940
Maghimaa M, Sagadevan S, Boojhana E, Fatimah I, Lett JA, Moharana S, Garg S, Al-Anber MA (2024) Enhancing biocompatibility and functionality: Carbon nanotube-polymer nanocomposites for improved biomedical applications. J Drug Deliv Sci Technol (2024) 105958.
Masenye ER, Mabuba N, Malinga SP (2022) Nanofiltration membrane based on hyperbranched polyethyleneimine-functionalised multiwalled carbon nanotubes for Pb (II) removal from water. Int J Environ Anal Chem 102(17):5601–5618
Mattausch H (2015) Properties and applications of nanoclay composites. Elsevier, Polymer nanoclay composites, pp 127–155
Metecan A, Cihanoğlu A, Altinkaya SA (2021) A positively charged loose nanofiltration membrane fabricated through complexing of alginate and polyethyleneimine with metal ions on the polyamideimide support for dye desalination. Chem Eng J 416:128946
Mohammad Gheimasi MH, Lorestani B, Kiani Sadr M, Cheraghi M, Emadzadeh D (2021) Synthesis of novel hybrid NF/FO nanocomposite membrane by incorporating black TiO 2 nanoparticles for highly efficient heavy metals removal. Int J Environ Res 15:475–485
Mondal R, De S (2022) Removal of copper (II) from aqueous solution using zinc oxide nanoparticle impregnated mixed matrix hollow fiber membrane. Environ Technol Innov 26:102300
Moon G-H, Kim D-H, Kim H-I, Bokare AD, Choi W (2014) Platinum-like behavior of reduced graphene oxide as a cocatalyst on TiO_2 for the efficient photocatalytic oxidation of arsenite. Environ Sci Technol Lett 1(2):185–190
Moradi M, Haghighi M, Allahyari S (2017) Precipitation dispersion of Ag–ZnO nanocatalyst over functionalized multiwall carbon nanotube used in degradation of Acid Orange from wastewater. Process Saf Environ Prot 107:414–427
Moud AA (2022) Polymer based flocculants: review of water purification applications. J Water Process Eng 48:102938
Mu W, Herrmann J-M, Pichat P (1989) Room temperature photocatalytic oxidation of liquid cyclohexane into cyclohexanone over neat and modified TiO_2. Catal Lett 3(1):73–84
Namdar H, Akbari A, Yegani R, Roghani-Mamaqani H (2021) Influence of aspartic acid functionalized graphene oxide presence in polyvinylchloride mixed matrix membranes on chromium removal from aqueous feed containing humic acid. J Environ Chem Eng 9(1):104685
Nanda T, Singh K, Shelly D, Mehta R (2021) Advancements in multi-scale filler reinforced epoxy nanocomposites for improved impact strength: a review. Crit Rev Solid State Mater Sci 46(4):281–329

Naushad M, Ahamad T, Al-Sheetan KM (2021) Development of a polymeric nanocomposite as a high performance adsorbent for Pb (II) removal from water medium: equilibrium, kinetic and antimicrobial activity. J Hazard Mater 407:124816

Nayak MC, Isloor AM, Lakshmi B, Marwani HM, Khan I (2020) Polyphenylsulfone/multiwalled carbon nanotubes mixed ultrafiltration membranes: fabrication, characterization and removal of heavy metals Pb2+, Hg2+, and Cd2+ from aqueous solutions. Arab J Chem 13(3):4661–4672

Neto VdOS, Freire TM, Saraiva GD, Muniz CR, Cunha MS, Fechine PBA, do Nascimento RF (2019) Water treatment devices based on zero-valent metal and metal oxide nanomaterials. In: Nanomaterials applications for environmental matrices, Elsevier, pp 187–225

Otrisal P, Obsel V, Buk J, Svorc L (2018) Preparation of filtration sorptive materials from nanofibers, bicofibers, and textile adsorbents without binders employment. Nanomaterials 8(8):564

Pan B, Xing B (2008) Adsorption mechanisms of organic chemicals on carbon nanotubes. Environ Sci Technol 42(24):9005–9013

Pandey S, Do JY, Kim J, Kang M (2020) Fast and highly efficient catalytic degradation of dyes using κ-carrageenan stabilized silver nanoparticles nanocatalyst. Carbohyd Polym 230:115597

Pandey RP, Ouda M, Abdul Rasheed P, Banat F, Hasan SW (2022) Surface decoration of bis-aminosilane cross-linked multiwall carbon nanotube ultrafiltration membrane for fast and efficient heavy metal removal. NPJ Clean Water 5(1):44

Peng X, Li Y, Luan Z, Di Z, Wang H, Tian B, Jia Z (2003) Adsorption of 1, 2-dichlorobenzene from water to carbon nanotubes. Chem Phys Lett 376(1–2):154–158

Phenrat T, Skácelová P, Petala E, Velosa A, Filip J (2020) Nanoscale zero-valent iron particles for water treatment: from basic principles to field-scale applications. Springer, Advanced nano-bio technologies for water and soil treatment, pp 19–52

Prabhu S, Poulose EK (2012) Silver nanoparticles: mechanism of antimicrobial action, synthesis, medical applications, and toxicity effects. Int Nano Lett 2(1):1–10

Qu X, Alvarez PJ, Li Q (2013) Applications of nanotechnology in water and wastewater treatment. Water Res 47(12):3931–3946

Raghunath A, Perumal E (2017) Metal oxide nanoparticles as antimicrobial agents: a promise for the future. Int J Antimicrob Agents 49(2):137–152

Ranjbaran H, Ameri E, Dehghani B (2024) Preparation of dendrimer/TiO2 polysulfone nanofiltration membrane to improve antibacterial, antifouling and separation performance of contaminants (heavy metals, salts, dyes). Polym Bull 81(2):1471–1494

Ren X, Chen C, Nagatsu M, Wang X (2011) Carbon nanotubes as adsorbents in environmental pollution management: a review. Chem Eng J 170(2–3):395–410

Sadegh H, Shahryari GR, Masjedi A, Mahmoodi Z, Kazemi M (2016) A review on Carbon nanotubes adsorbents for the removal of pollutants from aqueous solutions

Saeedi-Jurkuyeh A, Jafari AJ, Kalantary RR, Esrafili A (2020) A novel synthetic thin-film nanocomposite forward osmosis membrane modified by graphene oxide and polyethylene glycol for heavy metals removal from aqueous solutions. React Funct Polym 146:104397

Saifuddin N, Raziah A, Junizah A (2013) Carbon nanotubes: a review on structure and their interaction with proteins. J Chem

Santhosh C, Velmurugan V, Jacob G, Jeong SK, Grace AN, Bhatnagar A (2016) Role of nanomaterials in water treatment applications: a review. Chem Eng J 306:1116–1137

Seewoo BJ, Wong EV, Mulders YR, Goodes LM, Eroglu E, Brunner M, Gozt A, Toshniwal P, Symeonides C, Dunlop SA (2024) Impacts associated with the plastic polymers polycarbonate, polystyrene, polyvinyl chloride, and polybutadiene across their life cycle: a review. Heliyon

Shah V, Bhaliya J, Patel GM, Deshmukh K (2022) Advances in polymeric nanocomposites for automotive applications: a review. Polym Adv Technol 33(10):3023–3048

Sharma V, Singh H, Guleria S, Bhardwaj N, Puri S, Arya SK, Khatri M (2022) Application of superparamagnetic iron oxide nanoparticles (SPIONs) for heavy metal adsorption: a 10-year meta-analysis. Environ Nanotechnol Monitor Manag 18:100716

Sherugar P, Naik NS, Padaki M, Nayak V, Gangadharan A, Nadig AR, Déon S (2021) Fabrication of zinc doped aluminium oxide/polysulfone mixed matrix membranes for enhanced antifouling property and heavy metal removal. Chemosphere 275:130024

Sherugar P, Déon S, Nagaraja K, Padaki M (2022) Tailoring the structure of polysulfone nanocomposite membranes by incorporating iron oxide doped aluminium oxide for excellent separation performance and antifouling property. Environ Sci Water Res Technol 8(5):1059–1077

Shi C, Wei J, Jin Y, Kniel KE, Chiu PC (2012) Removal of viruses and bacteriophages from drinking water using zero-valent iron. Sep Purif Technol 84:72–78

Singh H, Bhardwaj N, Arya SK, Khatri M (2020) Environmental impacts of oil spills and their remediation by magnetic nanomaterials. Environ Nanotechnol Monit Manag 100305

Sreeramareddygari M, Shivanna JM, Somasundrum M, Soontarapa K, Surareungchai W (2021) Polythiocyanuric acid-functionalized MoS2 nanosheet-based high flux membranes for removal of toxic heavy metal ions and congo red. Chem Eng J 425:130592

Stefanakis AI, Becker JA (2016) A review of emerging contaminants in water: classification, sources, and potential risks. A Rev Emerg Contam Water Cllfication Sources Potential Risks 55–80

Surendhiran D, Sirajunnisa A, Tamilselvam K (2017) Silver–magnetic nanocomposites for water purification. Environ Chem Lett 15(3):367–386

Tan L, Xu J, Xue X, Lou Z, Zhu J, Baig SA, Xu X (2014) Multifunctional nanocomposite Fe 3 O 4@ SiO 2–mPD/SP for selective removal of Pb (ii) and Cr (vi) from aqueous solutions. RSC Adv 4(86):45920–45929

Tratnyek PG, Salter AJ, Nurmi JT, Sarathy V (2010) Environmental applications of zerovalent metals: iron vs. zinc. In: Nanoscale materials in chemistry: environmental applications. ACS Publications, pp 165–178

Üçel İS, Demirel E (2022) Modification of PVDF membranes using dopamine/zinc oxide for lead removal from aqueous media. Open J Nano 7(2):53–73

Uddin MT, Nicolas Y, Olivier C, Toupance T, Servant L, Muller MM, Kleebe H-J, Ziegler J, Jaegermann W (2012) Nanostructured SnO_2–ZnO heterojunction photocatalysts showing enhanced photocatalytic activity for the degradation of organic dyes. Inorg Chem 51(14):7764–7773

Umar K, Ibrahim MNM, Ahmad A, Rafatullah M (2019) Synthesis of Mn-doped TiO_2 by novel route and photocatalytic mineralization/intermediate studies of organic pollutants. Res Chem Intermed 45(5):2927–2945

Umesh AS, Puttaiahgowda YM, Thottathil S (2021) Enhanced adsorption: reviewing the potential of reinforcing polymers and hydrogels with nanomaterials for methylene blue dye removal. Surf Interfaces 104670

Vairavapandian D, Vichchulada P, Lay MD (2008) Preparation and modification of carbon nanotubes: review of recent advances and applications in catalysis and sensing. Anal Chim Acta 626(2):119–129

Wang K, Amin K, An Z, Cai Z, Chen H, Chen H, Dong Y, Feng X, Fu W, Gu J (2020) Advanced functional polymer materials. Mater Chem Front 4(7):1803–1915

WHO (2017) Safely managed drinking water: thematic report on drinking water 2017

Xiong Z, Lai B, Yang P, Zhou Y, Wang J, Fang S (2015) Comparative study on the reactivity of Fe/Cu bimetallic particles and zero valent iron (ZVI) under different conditions of N2, air or without aeration. J Hazard Mater 297:261–268

Yang M, Guo M, Xu E, Ren W, Wang D, Li S, Zhang S, Nan C-W, Shen Y (2024) Polymer nanocomposite dielectrics for capacitive energy storage. Nat Nanotechnol 19(5):588–603

Yaqoob AA, Ibrahim MNM (2019) A review article of nanoparticles; synthetic approaches and wastewater treatment methods. Int Res J Eng Technol 6:1–7

Zafar R, Zia KM, Tabasum S, Jabeen F, Noreen A, Zuber M (2016) Polysaccharide based bionanocomposites, properties and applications: a review. Int J Biol Macromol 92:1012–1024
Zaidi Z, Siddiqui SI, Fatima B, Chaudhry SA (2019) Synthesis of ZnO nanospheres for water treatment through adsorption and photocatalytic degradation: modelling and process optimization. Mater Res Bull 120:110584
Zak AK, Razali R, Abd Majid WH, Darroudi M (2011) Synthesis and characterization of a narrow size distribution of zinc oxide nanoparticles. Int J Nanomed 6:1399

Chapter 6
Thermoplastic Starch-Based Composites

Valiathan Sreejit, Radhakrishnan Preetha, and Gunasekaran Santhanalakshmi

Abstract Nowadays, researchers are mainly focusing on the fabrication of innovative biodegradable packaging materials. Since these materials are environmentally friendly, unlike petroleum-based plastics, starch forms one such economical biodegradable polymer. The granules of starch and their integral crystals are broken down into thermoplastic material with the excellent flowing capability to make a packaging material. Thermoformable starch is made from starch containing a higher concentration of amylose as it enables enhanced flow characteristics and a lesser probability of crystallization. Thermoplastic starch is manufactured using compression molding, extrusion, injection molding, thermoforming and reactive extrusion. This chapter discusses the different methods used for thermoplastic preparation and its large-scale production methods. When compared to plastics, the major defects of thermoplastic starch are its hydrophilicity and poor mechanical characteristics. The hydrophilic property of TPS can be reduced by the addition of other biodegradable polymers, cross-linking agents, or using the multi-layer technique. TPS reinforcement with cellulose microfibers, natural fibers, and industrially revived cellulose fibers will improve the mechanical characteristics of TPS. This chapter also provides an insight into the properties of Thermoplastic starch, Thermoplastic starch nanocomposites, characterization of Thermoplastic starch-based composites, biodegradability and, application of thermoplastic starch and its composites in detail.

Keywords Biodegradable · Thermoplastic · Starch · Extrusion · Nanocomposites

6.1 Introduction for Thermoplastic Starch

These days, research is concentrated on the development of biodegradable materials for packaging to overcome the problems resulting from large quantities of plastic disposal. Biopolymers are regarded as the primary substance used for this purpose,

V. Sreejit · R. Preetha (✉) · G. Santhanalakshmi
Department of Food Process Engineering, School of Bioengineering, Faculty of Science and Technology, SRM Institute of Science and Technology, Kattankulathur, Tamil Nadu, India
e-mail: preethar@srmist.edu.in

P. M. Visakh (ed.), *Engineering Applications of Polymer-Based Nanocomposites*, Advanced Structured Materials 237, https://doi.org/10.1007/978-3-031-93463-6_6

owing to their biodegradability and extended shelf life characteristics such as resistance to chemical or enzymatic reactions. Once a number of adjoined complexity levels are inferred in the structure, natural and artificial polymers become alike (Wunderlich 2011). Starch is one of the most excellent biopolymers that can be used for packaging. It is a universally available and low-priced sustainable product that can substitute traditional petroleum plastics when transformed into a thermoplastic. Due to the reasons mentioned above, investigating a large-scale process for converting starch into a thermoplastic substance became interesting research (Liu et al. 2009; Freitas et al. 2022). Starch is a conventional and economical biopolymer available in large quantities (Khan et al. 2017).

The major starch crops are maize, wheat, potatoes, rice, banana, cassava, yam, and sorghum (Carvalho 2008). Among these, some starches are converted to TPS (Chandra et al. 2023; Norrahim et al. 2024). The starch obtained from various sources has varying levels of amylose and amylopectin. The amylose content varies within the same botanical verity depending on geographic aspects and climatic conditions. It facilitates the synthesis of TPS films with different mechanical properties (Zhang et al. 2014a). Muscat et al. studied the outcome of using starch with different amylose content on TPS properties (Muscat et al. 2012).

The tensile strength and elastic modulus values were higher, and elongation at break was lower for films with a higher amount of amylose. However, factors such as intermolecular forces and hydrogen bonds in starch cause resistance and cause problems during thermoplastic preparation. The use of plasticizers can overcome this difficulty in obtaining thermoplastic starch. A plasticizer will enhance the flexibility and the process constancy of starch at a temperature below degradation temperature. The inclusion of a plasticizer will reduce the glass transition temperature (Khan et al. 2017). The gelatinization temperature of starch will increase due to fractional or complete water replacement by other plasticizers such as sugar and glycerol. This is due to the higher molecular weight of plasticizers other than water. It causes resistance to its penetration to the starch molecule. It will not be able to enhance the free volume of amorphous regions appreciably (Carvalho 2008). Starch is a natural polymer having composite levels of organization that impose upon thermoplastic distortion. The polymers that can flow as a result of heating beyond their melting temperature are known as thermoplastics. The main reasons underlying the plastic deformation of the polymers are its increased molar mass, enmeshment, interaction, and chain branches (Wunderlich 2011). Thermoplastic starch (TPS) is obtained due to alteration in the structure of the starch particle. This alteration is caused by processing the starch by applying temperature and shear force, along with a small quantity of water and plasticizing agents, which do not evaporate quickly during preparation (Perez-Pacheco et al. 2016). When starch is thermally processed, it can undergo a multiphase transition due to its distinctive microstructure. While the starch is thermally processed, water will diffuse into it, gelatinize, decompose, melt, and crystallize it. Thermoplastic starch is manufactured using conventional preparation methods, for instance, compression molding, extrusion, injection molding, thermoforming, and reactive extrusion (Robert and Ing 2012).

The granules of starch and their integral crystals should be disintegrated into thermoplastic material with the good flowing capability to make starch a packaging substance. The starch with a higher amount of amylose is chosen to synthesize thermoformable starch due to its improved flow characteristics and lesser probability of crystallizing. Corn starch is the most important raw material for obtaining amylose. An extruder is generally used to process the high amylose starch after mixing with water and plasticizers like glycerol. Water should disperse inside the granules and establish a hydrogen bond with the starch. Upon heating the starch granules, the granules experience destructurization, and the crystals melt resulting in the formation of homogenous viscous starch fluid (Robert and Ing 2012). Biodegradable polymers such as polyvinyl alcohol, polyoxyethylene, and polycaprolactone are incorporated with TPS to decrease TPS limitation, closely related to the hydrophilicity of TPS (Zhang et al. 2014a). After the starch film formation, the incorporated polymeric compound will even out the amorphous configuration by forming hydrogen bonds with the starch. It prevents the recrystallization of the starch from forming V-type crystals through the process of retrogradation. The retrogradation of starch can also be prevented by using substituted starch like hydroxyethyl, acetoxy starch. These substituted starches will disrupt the structural promptness of starch and hence limit the crystallization. There will be a considerable increase in the viscosity since the molecules are liberated from their regularly arranged superstructures. Hence this will result in the transformation of a gelatinous mass into a transparent one. The gelatinized starch will be amorphous, and it is known as TPS. The TPS can be processed repeatedly, provided the quantity of water is maintained or refilled. There is a characteristic glass transition temperature for TPS. It will flow on heating and crystallize gradually (Robert and Ing 2012). The sheets made out of TPS are generally transparent. However, depending upon the extent of shear blending, minute amounts of starch granules (ghost granules) are viewable as faint boundary lines. These granules cause a disturbance in the clarity of films/sheets and are not suitable for packaging purposes (Robert and Ing 2012). When compared to plastics, the major defects of thermoplastic starch are its hydrophilicity and poor mechanical characteristics (Juan et al. 2024). The hydrophilic property of TPS can be reduced to a particular extent by incorporating biodegradable polymers, cross-linking agents, or using a multi-layer technique. TPS reinforcement with cellulose microfibers, natural fibers, and industrially revived cellulose fibers will improve its mechanical characteristics (Ma et al. 2005). In a study reported by Nessi et al. (2019), it was seen that the incorporation of CNC to the thermoplastic starch improved the mechanical characteristics of the film. This chapter describes the properties, preparation, characterization, biodegradability, and application of thermoplastic starch and its composites.

6.2 Properties of Thermoplastic Starch

The barrier, mechanical, and thermal properties, retrogradation of starch, moisture sorption behavior, and surface tension of TPS are the factors influencing the utilization of TPS. These parameters are controlled by the source from which starch is obtained, the additives added such as processing factors, and the type of chemical treatment (Zhang et al. 2014b, Sharip et al. 2024).

6.2.1 Thermal Properties of Thermoplastic Starch

When starch is thermally processed such as kneading or extrusion, it undergoes several phase conversions such as glass transition, crystallization, gelatinization, and melting (Stavros et al. 2024). In the products prepared out of starch, the most important thermophysical property is glass transition (Tg).

As a result of applying glass transition temperature, the bond connecting the amylose fraction and amylopectin fractions of starch in the amorphous part of the starch is set free so that plasticity increases. The two fractions of starch gain adequate energy for large-scale molecular movement once the temperature rises above Tg. Whereas the extended range fragmental movement of the fractions of starch stops as soon as the temperature becomes less than Tg. The softness and flexibility of the polymer get reduced, and its stiffness and fragility increase, or vice versa during the glass transition. Once the temperature rises beyond Tg, there will be an abrupt change in physical characteristics such as increases in thermal expansion coefficient, heat capacity, dielectric coefficient, free volume, and visco-elastic characteristics. Tg is commonly determined using Differential Scanning Calorimetry (DSC) and dynamic mechanical analysis (DMA). DSC determines variations in the heat capacity of the TPS. The thermo-mechanical characteristics, such as loss modulus and storage modulus of TPS, are determined using DMA. In most cases, DSC will be less vulnerable to the glass transition of TPS compared to DMA. The native starch has a Tg of 243 °C following the Fox and Flory correlation. The incorporation of plasticizers results in the reduction of Tg of starch. The plasticizer has been proven to be highly efficient in reducing starch Tg (Zhang et al. 2014b). The glass transition temperature of various thermoplastic starches is given in Table 6.1.

6.2.2 Mechanical Properties of Thermoplastic Starch

Mechanical properties form the fundamental characteristics of TPS products. The mechanical properties of TPS, for instance, percentage of elongation at breakage (E), tensile strength (TS), modulus of elasticity (EM), and percentage of elongation at yield (EY), are determined using tensile tests. The tensile stress that a product

Table 6.1 Tg of various TPS

Thermoplastic starch formulation	Tg (°C)	References
Corn starch with triethyl citrate and glycerol	22 to 25	Zhang et al. (2011)
Pea starch with glycerol	66 to 75	Zhang and Han (2010)
Pea starch with sorbitol	33 to 47	Zhang and Han (2010)
Pea starch with glucose	26 to 49	Zhang and Han (2010)
Pea starch with fructose	19 to 31	Zhang and Han (2010)
Corn starch with glycerol, stearic acid, and water	10.3 to 12.3	Pushpadass and Hanna (2009)

made out of TPS can withstand before it breaks is given by TS value, the flexibility is depicted by the E value, and the stiffness by EM. Factors such as the sources from which starch is obtained, additives added, Tg, crystallinity, and the ratio of two fractions of starch influence the tensile characteristics of the TPS (Zhang et al. 2014b). The films made out of pure amylose are better than pure amylopectin concerning strength. TS increases with increasing concentration of amylose, and at the same time, TS decreases with an increase in amylopectin concentration. Films made out of high amylose corn starch exhibits improved tensile strength, elongation, impact strength, tear resistance, and steadiness at various relative humidity when judged against films with lesser amylose content. The starch films made up of different quantities of amylose and amylopectin will exhibit variations in mechanical properties. The variation in mechanical properties of the starch films is due to a greater degree of ordered structure in starch films containing higher quantities of amylose. The additives such as cross-linking agents, nanofillers, plasticizers, antioxidants, antimicrobial agents, and texturizing agents are used to regulate the tensile properties of TPS (Zhang et al. 2014b; Azra et al. 2023).

There will be a reduction in the value of TS and EM and an increase in the value of E to various levels as a result of the incorporation of plasticizer. Nevertheless, there will be a reduction in the E value if more amount plasticizer is used. This can be due to phase separation. For instance, the TPS sample will get separated into a starch-rich phase and plasticizer-rich phase once the glycerol content goes beyond critical concentration. This leads to a reduction in E value. It can also be due to the sliding of the added molecules of plasticizer. The strength of links between starch polymeric units becomes less, and there is a decrease in the cohesive energy of the polymer chains. Once the quantity of plasticizer rises beyond a critical concentration, it reduces E value (Zhang et al. 2014b). The mechanical properties of various thermoplastic starches are given in Table 6.2.

6.2.3 Barrier Properties of Thermoplastic Starch

Generally, all the polymers exhibit good relative permeability to small-sized molecules, for instance, O_2, CO_2, liquids, organic vapors, and water vapor. Based on

Table 6.2 Mechanical properties of different thermoplastic starch

Thermoplastic starch formulation	*TS (MPa)	**E (%)	***EM (MPa)	References
Banana starch oxidized with sodium hypochlorite	17.31 to 17.48	13.65 to 14.11	1.59 to 161	Alanís-López et al. (2011)
Rice starch film with sorbitol	11 to 22	3 to 4	456 to 1,053	Dias et al. (2010)
Cassava starch with glycerol	1.4 to 1.6	83 to 101	11 to 21	Müller et al. (2009)
Corn starch with glycerol and stearic acid	0.2 to 2.9	46 to 91	3 to 38	Pushpadass et al. (2009)
Corn starch with montmorillonite	1.91 to 2.96	5.26 to 6.26	51.8 to 61.4	Behera (2018)

the source of starch, plasticizer type, water hydration, quantity, and the temperature at which it is stored, the polymers offer a wide variety of mass transport properties from very high to small barrier values (Zhang et al. 2014b). The gas permeability coefficient (G) is used to express the gas barrier property. It specifies the amount of the gas that seeps into TPS packaging film per unit of area and time. The penetration of gas molecules generally occurs through the amorphous phase of starch, and it usually consists of absorption and diffusion. Initially, the gas molecules are taken up over the TPS film surface; afterward, the gas molecules will seep into the film. Then the gas molecules will disperse into the TPS film by Fick's first law. Permeation, diffusion, and absorption of gases are mass transfer processes in TPS packaging films (Zhang et al. 2014b).

G for a particular thermoplastic starch film is evaluated using Eq. 6.1.

$$G = TRg \times a/\Delta g \tag{6.1}$$

where a denotes the TPS film thickness, Δg denotes the variation among the partial pressure of gas across the film, and TRg denotes the gas transmission rate. Oxygen permeability (OP) and water vapor permeability (WVP) are employed to illustrate the gas barrier properties of the TPS film and describes the capacity of the TPS film to restrict the transportation of water vapor between a food product and the products nearby environment. The capability of the film to limit the transportation of water between the food item and its surrounding is measured as WVP. Generally, starch films are more hydrophilic than synthetic films; hence starch films have higher WVP. As a result, thermoplastic starch films are good barriers to water vapor. The inclusion of waxes, fatty acids, and neutral lipids can enhance the water barrier property of thermoplastic starch films. In work done by Dias et al. (2010), the WVP of TPS formulation with rice starch and glycerol was found to be 16.7 to 31.2 g mm m^{-2} h^{-1} kPa^{-1}, and with rice starch and sorbitol was found to be 9.6

to 10.9 g mm m^{-2} h^{-1} kPa^{-1}. In the research work performed by Pushpadass et al. (2009), the WVP of TPS formulation with corn starch and glycerol was found to be 16.7 to 31.2 g mm m^{-2} h^{-1} kPa^{-1}.

The oxygen barrier property is another significant factor that measures the gas barrier of the TPS film. TPS film generally has a significant barrier to O_2 when the moisture content is low, depending on its hydrophilic character. The O_2 permeability of thermoplastic starch film is slightly less than synthetic polymers such as ethylene–vinyl alcohol and low-density polyethylene. The plasticizing agent enhances the permeability of the film to O_2 by increasing the free volume and reducing crystallinity in the starch matrix (Zhang et al. 2014b).

6.2.4 Surface Tension of Thermoplastic Starch

The capability of thermoplastic starch to be printed, sealed, and coated is influenced by surface tension. Good adhesion is extremely wanted for sealing purposes of food packaging film. In the absence of sufficient adhesion, food may turn out to be contaminated with food pathogens. Surface tension also forms a vital factor for wettability. The high value of surface tension leads to the high-quality distribution of a liquid over the surface (Zhang et al. 2014b). The surface properties of TPS are affected by the ratio of amylose to amylopectin, pulsed light, chemical modifications, and plasticizers. In earlier studies, it was observed that amylopectin films are completely amorphous and provide smooth films. At the same time, amylose films are semicrystalline, leading to the formation of rougher films. The higher crystallinity of amylose films is due to the retrogradation of starch. The increase in amylose content of TPS films results in the following characteristics (Gutiérrez et al. 2018).

- The weak interaction between amylopectin and glycerol leads to a moisture-plasticizing effect.
- Reorganization of chains that increases crystallinity
- A more compact structure that brings about a reduction in water absorption
- The decrease in the smoothness of the surface that reduces its transparency
- Enhancement of contact angle that leads to improvement in the hydrophobicity of the film

In UV photo cross-linking, photoinitiators are added to the material. While processing, the presence of UV rays leads to the development of free radicals, which consequently start cross-linking reactions. The incorporation of sodium benzoate as a photosensitive agent in TPS/PVA under UV radiation results in improved hydrophobicity, tensile strength, and Young's modulus of the composite blend (Zhou et al. 2009). This was due to the enhancement in the cross-linking density of the macromolecular chains over the surface of the composite films (Gutiérrez et al. 2018). The pulsed light was used as a cross-linking agent in cassava film incorporated with taro starch and glycerol to develop food packaging material (Gutiérrez and González 2016). It was found that the pulsed light treatment resulted in the deterioration of the sample. Despite the amylose content of starch in the composite films, there was an increase in the crystallinity, contact angle, roughness, reduction in the moisture content, tensile strength, and transparency after treating it with pulsed light.

The properties of TPS films can also be varied through chemical modifications such as the addition of functional groups into the starch by acetylation, phosphation, oxidation, and so on. The above-said modifications bring about alterations in the molecular arrangement or establish functional groups and enhance starch usability. The addition of polar groups results in a smoother surface with more transparency and a smaller contact angle. At the same time, the incorporation of non-polar groups results in a rougher surface with less transparency and a larger contact angle (Gutiérrez et al. 2018). The addition of a plasticizer to the TPS blend results in a change in the mechanical and thermal properties of the blend. It brings down the rigidity at room temperature and reduces the temperature at which considerable deformations can be made using reasonable forces. It increases the elongation at break and stiffness of the film. The addition of plasticizer decreases the hydrophobicity of the film leading to lower moisture content and lower water solubility. The plasticizer also increases the flexibility of the film by reducing and enhancing elastic modulus, tensile strength, and strain at break. The incorporation of plasticizer also produces smoother films with more contact angle and transparency (Gutiérrez et al. 2018).

6.2.5 Property Variations of TPS

The structure of TPS varies with temperature, humidity, and time. It forms the major reason for the limited use of TPS in industrial applications. The hydrophilic nature of TPS results in a change in the moisture content of the starch at ambient humidity. Water content hysteresis depends on the increase or decrease of humidity. Moisture migration can occur between TPS and packed materials. It makes the packaged material or the TPS components soft. If water is absorbed into the packaged contents, the TPS components will be dried (Zhang et al. 2014b).

6.3 Preparation of Thermoplastic Starch and Thermoplastic Starch Nanocomposites

6.3.1 Preparation of Thermoplastic Starch

The TPS is mainly synthesized using extrusion (Dino et al. 2023). The other methods used for TPS synthesis are compression molding, injection moulding, and internal mixing. The primary technique used in processing TPS is extrusion, and it is generally in combination with various processes such as film casting, compression moulding, and injection moulding. One of the simplest techniques used for making starch films is film casting. However, it cannot be used for the commercial synthesis of starch (Xie et al. 2014a). Generally, two types of equipment are used for extruding the starch. They are single-screw extruding machines and twin-screw extruding machine (Xie et al. 2014b). The single-screw extruding machine can handle highly viscous starch and offer a high value of processing pressure for the uninterrupted production of unprocessed starch. A twin-screw extruding machine is comparatively useful for the initial step of TPS manufacturing since the twin-screw possesses the ability to self-wip. Also, a twin-screw extruding machine has considerable flexibility in its operation. It can control the temperature in each barrel zone. It has numerous sites for feeding the ingredients and flexible patterns for the diverse extent of blending. The twin-screw extruding machine helps good blending and compounding of the ingredients to form the TPS (Xie et al. 2014a).

The raw granules of starch are fed into the twin-screw extruding machine through a powder gravimetric feeder inside the extruding machine in a ravenous manner. The transmission capability of the screws placed in the feeding port is more than the feeding capacity. The liquid ingredients such as plasticizer are introduced mostly using a pressure injection for ensuring homogenous blending of the solid ingredients and liquid ingredients. Then the starch granules are subjected to elevated pressure and temperature inside the extruding machine. This will disrupt and gelatinize the starch granules. Once the TPS gets gelatinized, it is then melted and compressed (Xie et al. 2014a). The starch granules are degraded as a result of thermal and mechanical action inside the extruding machine. The amylopectin molecules are affected to a greater extent than amylose molecules. The fragmentation and degradation of the polymer chain occur selectively near the center of the starch molecule (Liu et al. 2010). This makes the size of the entities smaller and converges it to a maximum steady size. The TPS is passed along the extrusion line and eventually out from the die toward the end of the extruding machine, at a metered rate using a melt pump. The main purpose of the extruder is to ensure that the ingredients are completely melted and properly mixed to avoid degradation for maintaining high-quality mechanical traits. It results in the formation of uniform TPS melt. It can only be accomplished by understanding the effect of major processing factors such as the feed rate of raw materials, screw speed, temperature, and screw profile on extrudate quality (Xie et al. 2014a). One of the key factors that affect extrusion efficiency is the temperature profile. The temperature profile depends on the material fed to the extruder. Generally, initially, the starches are kept at a low temperature. The temperature rises to a maximum value at the central portion of the barrel, and then the temperature decreases to a certain level close to the die and at the die. The temperature at sections close to the feed port is kept low for preventing the loss of moisture once the untreated starch is introduced into the extruder. This is done because water is mostly used as a plasticizing agent. The temperature at the middle portion of the extruder is kept high to ensure strong thermo-mechanical action to destroy granules of starch and convert them into TPS. The application of low and suitable temperature at the die will facilitate the even extrusion of the feed material from the die. It prevents unwanted foaming in the presence of water, volatile materials, and so on. It also assists in orienting and crystallizing the material for improved characteristics, especially mechanical properties. The characteristics discussed above help determine residence time and pressure-specific mechanical energy input inside the extruder (Xie et al. 2014a).

The mechanical and rheological properties of the starch are influenced by the variation in the molecular weight of starch and the molecular weight distribution of starch. The degradation in the molecular weight of starch throughout the thermoplastic starch preparation is influenced by the temperature applied, the amount of plasticizer used, and the magnitude of specific mechanical energy used. The resulting product characteristics were used to study the influence of water on the extrusion of maize starch (Gomez and Aguilera 1984). They put forward a model in which the granules of starch are altered into gelatinized form. Subsequently, it was converted into free polymer chains, based on the extrusion parameters; it was then broken down into oligosaccharides or dextrinized starch or both and sugar molecules. This work mainly concluded that while the starch is extruded with less than 20% water content, a product distinctive from the gelatinized starch is achieved because starch is partly dextrinized. The alcohol precipitates the dextrin from the aqueous solution of starch. The dextrin is a dextrorotatory compound (Carvalho 2008). The degradation in molecular weight and melt rheology of amylopectin while the maize starch is extruded in a co-rotating twin-screw extruding machine was studied (Willett et al. 1997).

In the study conducted by Willett et al. 1997, the breakdown of starch molecules was measured with multi-angle light scattering (Willett et al. 1997). They observed a considerable reduction in molecular weight with the increase in the specific mechanical energy applied. It has been regarded as a critical factor in predicting the degradation in molecular weight of starch when it is extruded. How barely starch gets depolymerized when it extruded along with glycerol and water was studied by Myllymäki et al. (1997). They found a relationship between the quantity of glycerol and water used, starch chain depolymerization, and extruder screw speed. The starch granules are originally semi-crystalline. When the dry starch granules are heated, thermal breakdown takes place ahead of the melting point of the crystalline granule. Hence, the starch in its normal form cannot be subjected to melt-processing. The number of hydrogen linkages that keep the starch entities jointly must be brought down to subject the starch to melt-processing. The hydrogen linkages in starch can be reduced in a solvent, for example, water. The starch on heating in water, the phase transition occurs from a structured to an unstructured state and this course of action is known as gelatinization (Mohammadi Nafchi et al. 2013).

6.3.2 *Preparation of Thermoplastic Starch Nanocomposite Using Melt Mixing and Compression Molding*

The preparation of the thermoplastic starch nanocomposites involves two steps: melt mixing and compression molding. The steps involved in the preparation of thermoplastic starch nanocomposite are given in Fig. 6.1 (Ren et al. 2018).

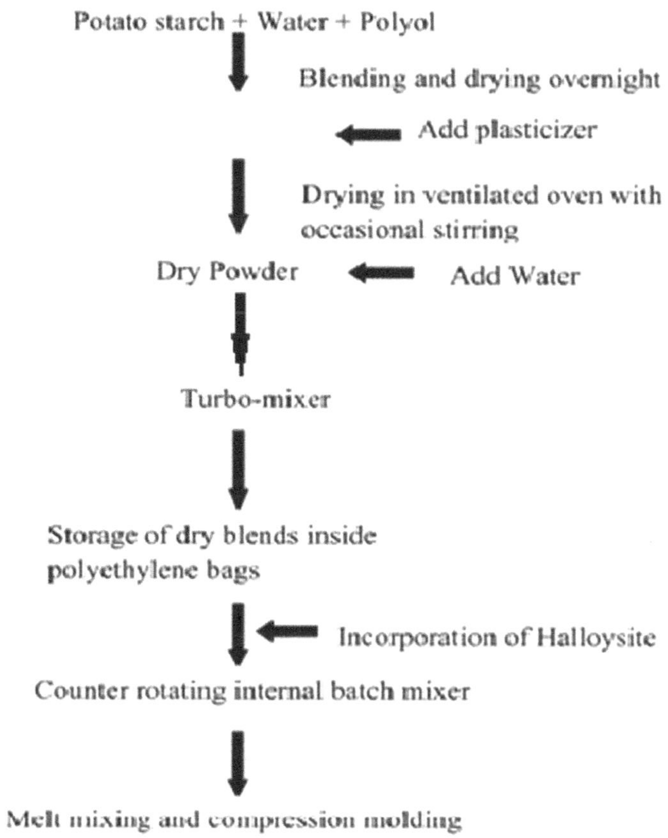

Fig. 6.1 Preparation of TPS nanocomposite using melt mixing and compression molding

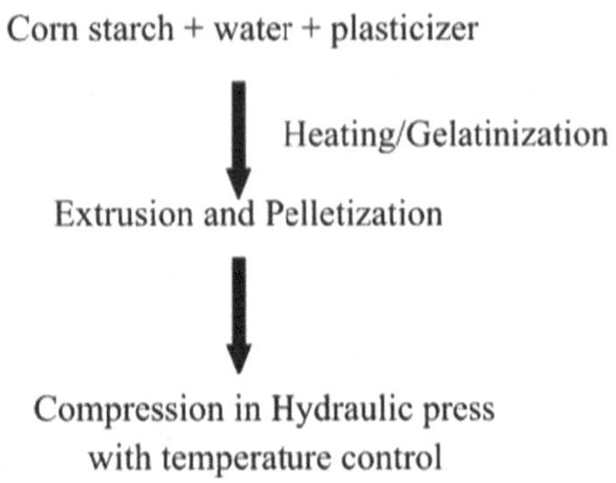

Fig. 6.2 Extrusion and compression of TPS nanocomposite

6.3.3 Extrusion and Compression of Thermoplastic Starch Nanocomposite

The thermoplastic starch films are manufactured using a three-step method: gelatinization, extrusion, and compression. The preparation procedure is given in Fig. 6.2 (González et al. 2020).

6.3.4 Injection Molding of Thermoplastic Starch

Injection molding is also reported for TPS nanocomposite preparation, and the procedure is given in Fig. 6.3 (Mondragón et al. 2009). The preparation and properties of the different TPS composites are given in Table 6.3.

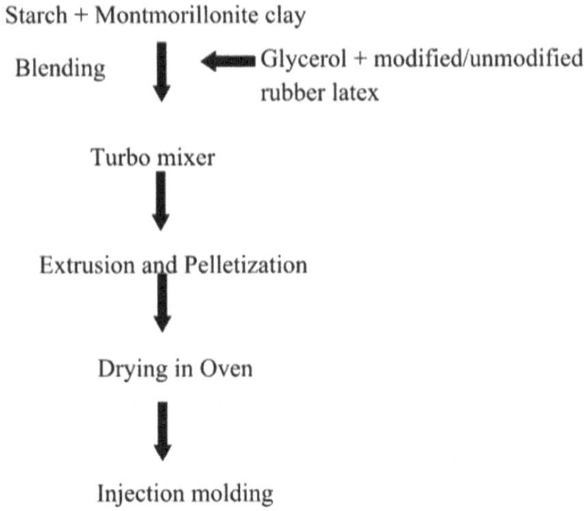

Fig. 6.3 Injection molding of TPS nanocomposite

Table 6.3 Preparation and properties of thermoplastic starch (TPS) composite

Starch and reinforcement fiber source	Composition of TPS composite	Properties of TPS composite	Reference
Cassava and Bagasse of Cassava	Starch was mixed with glycerol otherwise a combination of glycerol, stearic acid sorbitol, and various amounts of CNF obtained from cassava bagasse. They were compression molded into films	Significant reduction in the elastic module of theTPS composite film with 10 and 20% of CNF	Teixeira et al. (2009)
Potato and Potato tuber	A blend of microfibrils of cellulose was added to gelatinized solution of starch. It was plasticized using glycerol. The resulting blend was homogenized and molded in a teflon mold	The reinforcement increased the tension modulus of the film	Dufresne et al. (2000)
Potato and Wood flour	The starch was blended with stearic acid, sorbitol and varying amounts of CNF. The film was prepared using a twin-screw extruder and compression molding	The incorporation of CNF enhanced the tensile strength of the film	Hietala et al. (2013)
Barely grain and Husk of barely	The starch was mixed with water, glycerol, guar gum and cellulose fiber. The resulting blend was heated and it was transferred to acrylic plaques	The addition of CNF increased the tensile strength and brought down the elongation	Halal et al. (2015)
Corn and Eucalyptus wood	The film forming solution was prepared by blending starch, gelatin, glycerol, and the CNF and transferred to a Petri plate and dried	The CNF incorporated film showed better resistance	Freitas et al. (2014)
Wheat and Cotton	Starch, water and glycerol were blended to form a suspension, which was gelatinized followed by the incorporation of varying amounts of CNF This blend was degassed and poured over a mold made of polystyrene	The incorporation of CNF reduced the deformation and increased the Young's modulus and tensile strength of the film	Lu et al. (2005)
Chayote and Starch	The film forming solution was made with starch, water, glycerol and CNF. It was heated and transferred to petriplate and dried	The incorporation of CNF improved the mechanical property of the film	Aila-Suárez et al. (2013)
Cassava and Wood cellulose	Film forming solution was prepared with cassava starch, water, guar gum, glycerol and different amounts of CNF. Then molded in a petri plate and dried	The incorporation of CNF increased the tensile strength and reduced WVP and deformation	Müller et al. (2009)

(continued)

Table 6.3 (continued)

Starch and reinforcement fiber source	Composition of TPS composite	Properties of TPS composite	Reference
Tapioca and Sesbania Javanica	The blend was prepared by blending tapioca starch, water, glycerol and CNF. The blend was melt compounded in a twin screw extruder and then it was compression molded into films	The addition of CNF significantly improved the moisture resistance and modulus	Kampangkaew et al. (2014)
Rice and Cotton fiber	The film forming solution was made by mixing starch, glycerol, stearic acid and different amounts of cotton fiber. It was then compressed into thick films	The addition of cotton fiber in the starch film improves its tensile strength and reduced its hydrophilicity	Prachayawarakorn and Pattanasin (2016)
Corn and Sisal	The blend was prepared by mixing corn starch, water, glycerol and different concentrations of cellulose nanofiber. It was then compression molded into film	The incorporation of CNF improved the mechanical properties of the film	Orue et al. (2016)
Corn and Wheat straw	TPS corn starch was blended with liquid glycerol and stored overnight. Then an internal mixer was used for melt-blending the starch sample After cooling the TPS, different quantities of CNF were added and blended in an internal mixer and extruded along with polyethylene	The increase in the amount of CNF incorporated in the film resulted in the increase of Young's modulus	Alidadi-Shamsabadi et al. (2015)

6.4 Scaling up of TPS and TPS Nanocomposite Manufacturing

The TPS is synthesized using the casting process or extrusion process. The first step in the casting process is to formulate starch by adding a plasticizing agent and other additives into a starch water suspension, and then gelatinized by heating. It is cast on a plane surface, cooled, dried, and the starch film is peeled from the plane surface. The quality of TPS is influenced by the duration of heating and the temperature used. In work done by Koch et al. (2010), it was observed that the heating of starch at 130–150 °C for 5–15 min resulted in the breakdown of starch and formation of disjointed TPS films. Compared to heating, the temperature used for the heating process had an influence on the breakdown of starch to a greater extent (Zhang et al. 2014a). Generally, the casting process is applied on a lab-scale for studying the gelatinization, plasticization, and thermo-mechanical characteristics of the TPS films. On the commercial scale, extrusion is used. A single-screw extruder is used for carrying out extrusion. The different parts of an extruder are hopper, barrel, feed screw, thermocouples, and dies. Starch pellets are made using the extruder. The heat energy and shear stress will destroy the starch crystallinity. It will result in the development of continuous amorphous mass (Mościcki et al. 2012). The various constituents present in the starch blend such as water and plasticizing agents interact with the starch through hydrogen bonding. Then the starch undergoes fragmentation, melting, and flowing similar to artificial polymers.

The starch pellets are mixed with different additives like artificial polymer or else nanoclay, and it is introduced into the extruder. It initiates from the hopper and then passes beside the feed screw in the barrel compartment. The pellets of starch undergo uniform melting and mixing with the additives. It is then moving to the screw die. The die is accurately designed with a particular aperture such that the cross-section of the product coming out from the extruder will have the pattern of the die. The product coming out from the die will solidify rapidly. Hence, the TPS product from the extruder is either blown into sheets or films or molded to form the desired pattern (Zhang et al. 2014a). The influence of extrusion and casting on the characteristics of TPS blend films was studied by Melo et al. (2011). It was concluded that casting gives rise to transparent, uniform, and smooth films. In addition to that, casting provided films with low water vapor permeability compared to the films prepared by extrusion, which resulted in opaque films.

6.5 Thermoplastic Starch-Based Composites and Their Characterization

6.5.1 Morphological Characterization

The morphological characterization of thermoplastic Starch-Based Composites is generally done using scanning electron microscopy (SEM), transmission electron microscopy (TEM), and atomic force microscopy (AFM). The SEM analysis was used to understand the morphology of the composite. There are many reports on the dimensions and shapes of starch granules using SEM. The TEM analysis was also used to understand the size distribution and morphology of the composite (Shruthy and Preetha 2019). The AFM was used to study the surface quality and distribution of filler materials in the composite blend (Julinová et al. 2019). According to the morphological characterization results, the starch granules occur in different shapes. Total destruction of its structure and the alteration into thermoplastic starch is evident on the micrographs of the blends of thermoplastic starch and other ingredients. Starch granules occur in various shapes and sizes based on the source from which it was obtained. The starch granules may be oval, elliptical, circular, or polygonal (Azevedo et al. 2019). In the study performed by Bulatovic et al. (2019), the SEM examination showed that the surface of the composite of Polycaprolactone (PCL) with 30 weight %

TPS was smooth with a small surface roughness that was caused due to imperfectly plasticized starch. The smooth surfaces are characteristic of the PCL matrix that degrades at a slower rate, and sharp holes are characteristic of the thermoplastic starch phase that degrades faster.

The TEM analyses were performed in the research work done by Behera (2018), and it showed that in starch montmorillonite composites without nanoclay, a smooth surface of thermoplastic starch is present. In contrast, in starch montmorillonite composites with 3% nanoclay, clay layers could be noticeably seen, organized layer by layer representing intercalation of silicate layers. In the composite made of TPS Cloisite 30B with 3% nanoclay, each clay layer is detached, indicating the exfoliation of nanoclay in thermoplastic starch. The intercalation and exfoliation of nanoclay assist in enhancing the mechanical characteristics of TPS nanoclay composite. The biodegradation rate of a composite depends on the topographical features of the material and the nature of the polymer, stabilizer, and plasticizing agent used. The progression of microbes in the composite is affected by the roughness of the composite material. In the study done by Julinová et al. (2019), the AFM images showed that the unevenness on the surface of TPS and its blends. The incorporation of fillers did not remarkably influence the biodegradation of the TPS composite.

6.5.2 *Optical Properties*

The UV–Visible spectroscopy is used for the determination of transparency of thermoplastic starch composites. As the opacity decreases, the transparency increases (Shruthy and Preetha 2019). The transparency can be an indirect measurement of the amount and distribution of cellulose nanofibers inside the composite (Hietala et al. 2013).

6.5.3 *Microstructure and Crystalline Studies*

The microstructure and crystalline properties of thermoplastic starch composites are inferred with the help of X-ray diffraction. It is also used to study the changes in the configuration of the starch once it has plasticized. The study of the corn starch (CS) showed a high proportion of crystallinity of A-type and diffraction peaks of CS were positioned at $2\theta = 15.1°$, 18°, and 23.1° (Wang et al. 2015). X-ray diffractograms revealed the variation in the crystalline structure of starch after treatment with a formamide-urea. The crystalline arrangement seen in the CS granules are shattered in thermoplastic starch; the development of B and V-type crystalline structures took place due to the recrystallization of starch. Hence, it was shown that the plasticizer molecule blocked the formation of the hydrogen bonds connecting the starch molecules and produced new hydrogen bonds among molecules of starch (Wang et al. 2015). In another study done by Azevedo et al. (Teixeira et al. 2009), X-ray diffraction depicted the crystalline arrangement of the thermoplastic starch composite. There was a strong linkage between whey protein isolate and thermoplastic corn starch due to the amorphous nature of whey protein isolate. Two crystalline configurations should be seen for the thermoplastic starch composite with peaks at approximately $2\theta = 13.4^\circ$ and $2\theta = 20^\circ$, consequent to the VH type crystalline arrangement crystallization of amylose. It also showed B-type crystallinity having a peak at approximately $2\theta = 17.4^\circ$, connected with the amylopectin recrystallization.

6.5.4 Thermal Properties

The thermal properties of the TPS composites are studied using Thermogravimetric analysis (TGA) and Differential scanning calorimetry (DSC). The TGA is used to infer the quantity and rate of variation in the weight of a substance with respect to time and temperature in a controlled environment (Zain et al. 2019). The DSC is used to study the differences in the thermal properties of composites (Zain et al. 2019). The proper understanding of decomposition and mode of degradation under the action of heat is essential for the standardization of thermoplastic starch processing. The threshold decomposition temperature can provide the specification of the maximum processing temperature (Mano et al. 2003).

Moreover, from the thermogravimetric analysis, it was observed that the incorporation of nanocomposites prevents the degradation of TPS at low temperatures. For example, the incorporation of montmorillonite clay enhanced the stability of TPS at high temperature (Azevedo et al. 2019). In another research work, the TGA curve of the composite showed three stages of degradation. Initially, there was degradation in a temperature range from 90–200 °C, which was characteristic of the evaporation and removal of glycerol and water in the TPS. It was followed by degradation at a temperature of 250–350 °C was characteristic of degradation of TPS present in the composite. Finally, the degradation of low linear density polyethylene was shown at 380–480 °C (Zain et al. 2019). In another study the DSC showed a drop in the polycaprolactone (PCL) melting peak due to the addition of TPS to the PCL matrix (Bulatović et al. 2019). This was due to the decrease in the chemical potential of the polymer as a result of the occurrence of a moderately miscible amorphous polymer. TPS facilitates the cold crystallization of PLA faster because TPS acts as a nucleating agent and triggers the cold crystallization of PLA. Hence there will be a considerable gain in the crystallinity of PLA.

6.5.5 Mechanical Properties

The use of thermoplastic starch is feasible for preparing a wide variety of products such as compostable garbage bags, protective foam, single usage bags, etc. One of the most important factors to be considered while selecting TPS for such an application is its mechanical characteristics. TPS should be able to compete with conventional plastic with respect to their mechanical properties and processability. The mechanical properties such as elongation property and tensile strength are determined using a digital tensile testing machine. In the research work done by Azevedo et al. (2019), the incorporation of montmorillonite to the composite of TPS and whey protein isolate (WPI) enhanced the elastic modulus and subsequently decreased the material flexibility and elongation. The scattering of montmorillonite nanoparticle reinforcement occurs due to the high surface area and aspect ratio of the silicate layers. Hence it forms a product with good dispersal of the clay stratum inside the matrix of WPI. The strong connection within the clay stratum and the WPI enhances elastic modulus (Azevedo et al. 2019).

In work done by Sessini et al. (2019), TPS exhibited the least elongation at break (E), and Ethyl vinyl acetate (EVA) had the maximum value of E when judged against the other samples. As expected for elastomers, a rather lesser value of elastic modulus was seen for EVA. TPS exhibited a characteristic elastic modulus of an amorphous polymer in the rubbery condition since at room temperature. With the enhancement in the TPS content inside the blend, the substances became more delicate, resulting in a higher elastic modulus. It also brings down tensile stress compared to pure EVA. Fascinatingly, the nanocomposites had lower elastic modulus in comparison with the equivalent neat blends.

6.5.6 Water Vapor Permeability (WVP)

To obtain superior freshness and preserve food items, it is particularly essential to prevent or reduce moisture movement between the food and its atmosphere. Hence, being a barrier property, the WVP is vital and must be of minimum value for food packaging to enhance the shelf life of the food products. The chemical/physical degeneration of food is controlled by moisture content, and moisture content is controlled by WVP of the packaging film (Castillo et al. 2017). Many studies reported that the TPS nanocomposites have the minimum WVP rate compared to pure TPS packaging film (Bulatović et al. 2019).

6.5.7 Molecular Interaction Study Using Fourier Transform Infrared Spectroscopy

Electromagnetic waves that interact with the material will undergo absorption, transmission, reflection, scattering, or possess photoluminescence, which gives important details about the molecular organization and the energy level change of that material (Munajad and Subroto 2018). In work done by Chen et al. (2019), the FTIR depicted the hydrogen bonding between the graphene quantum dots (GQD) and TPS. The absorption band at 3200 to 3500 cm^{-1} in the FTIR spectrum corresponds to intermolecular and intramolecular hydrogen-bonded hydroxyl groups in the starch and glycerol. The above result indicated that the hydrogen bonds were not affected by incorporating a lower concentration of GQD. In another study, the incorporation of lignin-containing cellulose nanofibrils (LCNF) to TPS resulted in reduced frequency to 3265 cm^{-1} from a value of 3292 cm^{-1}. It was due to the development of hydrogen bonds in the composite film. This makes TPS and LCNF more compatible and reduces its water vapor permeability (Zhang et al. 2020).

6.6 Biodegradation Studies of Thermoplastic Starch-Based Composites

The biodegradation of waste mainly takes place through anaerobic digestion at a temperature of 37 ºC. The application of TPS based packaging material can be used to solve the problem caused due to the non-biodegradable synthetic packaging materials (Bulatović et al. 2019). The extent of biodegradation of a polymer is affected by microorganisms, water availability, O_2 availability, chemical atmosphere, and temperature (Julinová et al. 2019). The microorganisms that play a major role in the degradation of biodegradable organic matter are aerobic mesophilic and thermophilic microorganisms. There will be an increase in the composting mass temperature due to the release of a large amount of heat. There are four phases in the composting process: mesophilic, thermophilic, cooled, and maturation phase. The most critical phase is the thermophilic phase, at which a high value of temperature facilitates biodegradation by microorganisms—the biodegradation results in the reduction of weight of the composite. The hydrophilic nature of starch makes its degradation easier. The TPS is easily attracted by microorganisms and available to them, hence facilitates the biodegradation process. In a study of TPS, the mixture with a higher concentration of Polycaprolactone (PCL) exhibited retarded rate of biodegradation compared to the mixture with a low concentration of PCL due to its crystalline structure and hydrophobicity (Bulatović et al. 2019). Two factors that are typically used to check the extent of biodegradation measure the weight loss of TPS film over time. The second method is finding out the quantity of CO_2 released during composting. In the literature, the biodegradation ability of the TPS composite in an aqueous medium was reported in the O_2 rich atmosphere. Activated sludge from municipal effluent treating plants containing mixed microbial culture was used for conducting this study. The biodegradability of TPS composite in an aqueous medium was expressed as the ratio of CO_2 originally formed during biodegradation to the hypothetical production of CO_2 from the substratum as given in Eq. 6.2 (Julinová et al. 2019). The incorporation of clay dispersible filler to the TPS matrix supported the biodegradation to a greater extent than water non-dispersible fillers.

$$\mathrm{R_{CO2}} = (Qs - Qb/Tp) \times 100 \tag{6.2}$$

where R_{CO2} is the remaining quantity of carbon in the test material expressed as %,

Qs is the amount of CO_2 formed as a result of degradation of test material,

Qb is the amount of CO_2 formed as a result of endogenous respiration,

Tp is the amount of CO_2 formed hypothetically due to the total degradation of the test material.

The rate of biodegradation of TPS composite was also studied using its biodegradation in soil. Here an electrochemical respirometer was used to determine Biological Oxygen Demand (BOD) for evaluating the rate of biodegradation. The test was done until a steady BOD level was attained, and no more biological degradation was anticipated. The quality of biodegradation of the TPS composite was calculated using Eq. 3 and expressed as a percentage.

$$R_{BOD} = [(BOD_s - BOD_b)/Ws * COD_s] \times 100 \quad (6.3)$$

BOD_b denotes the BOD of blank in mg, BODs denote the BOD of the sample during the biodegradation of the sample, CODs is the Chemical Oxygen Demand of the sample in mg g^{-1}, and Ws is the weight of the sample. The TPS composite incorporated with clay fillers exhibited the maximum biodegradation rate, and the TPS composite containing hydrophobized wastepaper filler showed the least biodegradation (Julinová et al. 2019).

6.7 Applications of TPS and Thermoplastic Starch-Based Composites

The TPS and its composites are used in various fields such as food technology, biomedical engineering, environmental engineering, and so on.

6.7.1 Applications in Food Technology

The primary role of packaging is to make the product cost-effective and offer protection from environmental factors. Starch is one of the major materials being chosen to make such packaging material. However, plasticizers are incorporated to prevent the retrogradation of starch. It improves the processing ability of the starch (Khan et al. 2017). The TPS has proved to be a promising material that can replace petroleum-based packaging materials that are not easily degradable. TPS can be used as a replacement for polystyrene (Jabeen et al. 2015).

TPS is reinforced using either natural organic fillers or mineral fillers to reduce its moisture vulnerability and improve its mechanical strength. Food packaging bags were developed from thermo-compressed films made from thermoplastic corn starch and talc nanoparticles. The incorporation of talc nanoparticles made the bags stronger

Table 6.4 Application of thermoplastic starch-based composites

Source of starch	Reinforcing agent	Possible application	References
Cassava	Polyvinyl alcohol	Packing of edible products	Baratter et al. (2017)
Cassava	Cotton fiber, pectin	Packing of food	Prachayawarakorn and Pattanasin (2016)
Cassava	Cellullose nanofiber	Biodegradable substance	Fahma et al. (2017)
Cassava	Silica, Cellulose pulp	Biodegradable substance	Raabe et al. (2015)
Sugar palm	Seaweed	Packaging	Jumaidin et al. (2016)

(López et al. 2015). In short, the incorporation of various fillers can improve the strength, hydrophobicity, and biodegradability of thermoplastic starch (Zélia et al. 2024). The TPS composites have been used for food packaging, fabrication of edible films, and biodegradable packaging; examples are given in Table 6.4.

6.7.2 Applications in Biomedical Engineering

The TPS polymers are used as matrices to deliver drugs at a controlled rate. For instance, microparticles made from polyacrylate modified TPS are used as a carrier for drugs. Microparticles made using albumin and TPS are fabricated to release the protein at a controlled rate. Research work has also been done for the TPS based controlled release scaffolds for releasing growth factors. TPS polymers are used in the field of tissue engineering for controlling the rate of drug release. The blends of starch polylactide and starch polyethylene, polyvinyl alcohol, are used to substitute the biomedical implants (Halley et al. 2007). The TPS-Polyvinyl alcohol blend that is processed in the extruder is used to prepare capsules for drug delivery. Generally, the TPS is chemically modified at a controlled pace for enabling it to be used for the required purpose. The different types of chemical modifications typically done on starch are acetylation, carboxymethylation, succinylation, hydroxypropylation, and phosphorylation. The substitution extent relies upon the reaction time, pH, concentration of reactant, molecular organization of substituent molecule, and the morphology of the granules. The TPS is used for the synthesis of scaffolds. Scaffolds are defined as three-dimensional arrangement designed to support the growth of the cells (Carvalho and Trovatti 2016). The material used for the preparation of the scaffold should have the required mechanical features for facilitating the adhesion of cells and tissue formation. It should also possess a satisfactory structural design for the regeneration of the tissues. Apart from these factors, the size of pore, permeability, and porosity also affects the design of the scaffold. Based on the tissue and its role, the scaffold should break down, and the disintegrated product should be taken in at different periods of time, generally three months. It is the time needed for the development of nearly all of the tissues. Various techniques are being developed for bringing together starch scaffold structures, such as fiber bonding and extrusion.

The hierarchical arrangement of scaffolds impersonates the indigenous extracellular matrix and enhances the biological functioning of the substance that facilitates the multiplication of cells that resemble osteoblast. Three-dimensional printing is also being successfully used for the preparation of starch scaffolds (Qing et al. 2022). The scaffold prepared from a composite of dextran, corn starch, and gelatin exhibited good mechanical characteristics (Zhang et al. 2020).

In work done by Mahdieh et al. (2016), a nanocomposite material was made from TPS, Ethyl vinyl alcohol, vitamin E, and nanoforsterite using extrusion and injection moulding. This nanocomposite had required mechanical energy to be used as a bone scaffold and for other orthopedic purposes.

6.7.3 Applications in Environmental Technology

The demand for packaging material with biodegradability is increasing these days significantly. Hence packaging materials with efficient biodegradability are being developed using biodegradable polymers such as starch. When these polymers are disposed to the environment, such as landfills, microbial organisms are broken down through enzymatic catalysis. It results in the formation of CH_4, CO_2, biomass, water, and so on (Lörcks 1998). Bioplastics are biodegradable polymers that can be disposed of to the environment without causing any problems to the environment. The important features of degradable bioplastics made from TPS are.

- it can be processed using the available plastic manufacturing industries
- it has a high permeability to water vapor
- it has excellent O_2 permeability
- it is not chargeable electro-statically

Bioplastics are used for the preparation of degradable packaging, temporary consumable items, and unique goods. The blown films made from the TPS are used to manufacture mulch foils, trash bags, protective clothing, air bubble films, gloves, barrier ribbons, labels, and so on. The flat films made from the TPS are used to manufacture trays for food and non-food products, packaging for frozen products, pharmaceutical packaging, cups, and so on. The TPS is injection molded to manufacture toys, golf tees, disposable cutlery, cans, writing materials, and so on (Lörcks 1998). A strategy to reduce the price of biodegradable polymeric substances is to mix the biodegradable polymers with TPS. Starch is susceptible to water, and the characteristics of films developed using starch are affected by the amount of moisture. Hence, it exhibits comparatively lower mechanical resistance.

Furthermore, the starch does not possess good thermal processability due to its lesser thermal decomposition temperature and higher melting point. Therefore, TPS is blended with polymers such as polyhydroxyalkanoates, polycaprolactone, polylactic acid, and polyolefins to obtain biodegradable and economical packaging materials. TPS blends with biodegradable polymers are processed using extrusion, film blowing, blow molding, foaming, and injection molding. The polypropylene,

Table 6.5 Commercially used TPS polymers

Name of the product	Material used	Application
ECOFOAM	Tapioca starch and modified maize starch	Extruded sheets, injection molded toys, expanded foams
MATERBI	TPS polymer blends	Thermoformed commodities, injection molded products, biodegradable mulch bags and films
SOLANYL	Modified potato starch	Injection molded commodities
BIOPLAST	TPS	Cups, boxes, bags, horticultural films, cutlery, tableware, planting pots, wrap films and trays
EARTHSHELL PACKAGING	Starch from tapioca, potatoes, wheat, corn and rice together with fiber, water, air and biodegradable micro-thin coatings	Foamed plates, trays, packaging laminates and films
PLANTIC	TPS using corn starch	Thermoformed trays

belonging to the polyolefins group, is made more compostable by compounding it with starch and processing using extrusion and injection molding. Compatibilizer such as maleic anhydride is added for enhancing the distribution of starch in polypropylene matrix and for improved linkage between starch and polypropylene.

6.7.4 Commercial Applications

The commercial application of TPC is reported for manufacturing foams, toys, bags, cups, cutleries, pots, trays, and so on. Details of commercial products are given in Table 6.5 (Halley et al. 2007).

6.8 Conclusion

The exercise of using synthetic packaging materials are ultimately resulting in the contamination of the environment. Hence, it is a necessity to do research and create novel packaging materials that are eco-friendly. The development of biodegradable composites with thermoplastic starch can be one of the methods used for accomplishing this purpose. However, it is required to use fillers or reinforcement to enhance the flexibility, strength, and barrier properties of the thermoplastic starch for better utilization. This chapter discusses the preparation, properties, characterization, biodegradability, and applications of thermoplastic s and its composites. The details furnished in this chapter can help give an insight into the various aspects required for the development of thermoplastic starch-based composites.

References

Aila-Suárez S, Palma-Rodríguez HM, Rodríguez-Hernández AI, Hernández-Uribe JP, Bello-Pérez LA, Vargas-Torres A (2013) Characterization of films made with chayote tuber and potato starches blending with cellulose nanoparticles. Carbohyd Polym 98(1):102–107

Alanís-López P, Pérez-González J, Rendón-Villalobos R, Jiménez-Pérez A, Solorza-Feria J (2011) Extrusion and characterization of thermoplastic starch sheets from "macho" banana. J Food Sci 76(6):E465–E471

Alidadi-Shamsabadi M, Behzad T, Bagheri R, Nari-Nasrabadi B (2015) Preparation and characterization of low-density polyethylene/thermoplastic starch composites reinforced by cellulose nanofibers. Polym Compos 36(12):2309–2316

Azevedo VM, Carvalho RA, Borges SV, Claro PIC, Hasegawa FK, Yoshida MI, Marconcini JM (2019) Thermoplastic starch/whey protein isolate/rosemary essential oil nanocomposites obtained by extrusion process: antioxidant polymers. J Appl Polym Sci 136(23):47619

Azra NA, Atiqah A, Jalar A, Manar G, Supian ABM, Ilyas RA (2023) Sustainable substrate tin oxide/nanofibril cellulose/thermoplastic starch: dimensional stability and tensile properties. J Market Res 26:99–108

Baratter M, Weschenlfelder EF, Stoffel F, Zeni M, Piemolini-barreto LT (2017) Analysis and evaluation of cassava starch-based biodegradable trays as an alternative packaging to fresh strawberry (Fragariaananassa cv San Andreas) 3(4):76–81

Behera AK (2018) Mechanical and biodegradation analysis of thermoplastic starch reinforced nanobiocomposites. In: IOP conference series: materials science and engineering, vol 410, no 1. IOP Publishing, p 012001

Bulatović VO, Grgić DK, Slouf M, Ostafinska A, Dybal J, Jozinović A (2019) Biodegradability of blends based on aliphatic polyester and thermoplastic starch. Chem Pap 73(5):1121–1134

Carvalho AJ (2008) Starch: major sources, properties and applications as thermoplastic materials. In: Monomers, polymers and composites from renewable resources. Elsevier, pp 321–342

de Carvalho AJF, Trovatti E (2016) Biomedical applications for thermoplastic starch. Biodegrad Biobased Polym Environ Biomed Appl 1–23

Castillo LA, López OV, García MA, Villar MA, Barbosa SE (2017) Biodegradable composites based on thermoplastic starch and talc nanoparticles. Handb Compos Renew Mater 5:23–59

Chandra MC, Harini K, Stefano F, Diego R (2023) Development and characterization of starch-based bioactive thermoplastic packaging films derived from banana peels. Carbohydr Polym Technol Appl 5

Chen J, Long Z, Wang S, Meng Y, Zhang G, Nie S (2019) Biodegradable blends of graphene quantum dots and thermoplastic starch with solid-state photoluminescent and conductive properties. Int J Biol Macromol 139:367–437

Dias AB, Müller CM, Larotonda FD, Laurindo JB (2010) Biodegradable films based on rice starch and rice flour. J Cereal Sci 51(2):213–219

Dino R, Emma MM, Joo ST, Chee KL (2023) Thermoplastic starch (TPS) bioplastic, the green solution for single-use petroleum plastic food packaging–a review. Enzym Microb Technol 168

Dufresne A, Dupeyre D, Vignon MR (2000) Cellulose microfibrils from potato tuber cells: processing and characterization of starch–cellulose microfibril composites. J Appl Polym Sci 76(14):2080–2092

El Halal SLM, Colussi R, Deon VG, Pinto VZ, Villanova FA, Carreño NLV, da Rosa Zavareze E (2015) Films based on oxidized starch and cellulose from barley. Carbohyd Polym 133:644–653

Fahma F, Sunarti TC, Indriyani SM, Lisdayana N (2017) Thermoplastic cassava starch-pva composite films with cellulose nanofibers from oil palm empty fruit bunches as reinforcement agent, pp 1–6
de Freitas A de SM, da Silva APB, Montagna LS, Nogueira IA, Carvalho NK, de Faria VS, Lemes AP (2022) Thermoplastic starch nanocomposites: sources, production and applications–a review. J Biomater Sci Polym Ed 33(7):900–945
Freitas F, Alves VD, Reis MA, Crespo JG, Coelhoso IM (2014) Microbial polysaccharide-based membranes: current and future applications. J Appl Polym Sci 131(6)
Gomez MH, Aguilera JM (1984) A physicochemical model for extrusion of corn starch. J Food Sci 49(1):40–43
González K, Iturriaga L, González A, Eceiza A, Gabilondo N (2020) Improving mechanical and barrier properties of thermoplastic starch and polysaccharide nanocrystals nanocomposites. Eur Polymer J 123:109415
Gutiérrez TJ, González G (2016) Effects of exposure to pulsed light on surface and structural properties of edible films made from cassava and taro starch. Food Bioprocess Technol 9(11):1812–1824
Gutiérrez TJ, Ollier R, Alvarez VA (2018) Surface properties of thermoplastic starch materials reinforced with natural fillers. Editorial Springer International Publishing, Functional biopolymers, pp 131–158
Halley PJ, Truss RW, Markotsis MG, Chaleat C, Russo M, Sargent AL, Tan I, Sopade PA (2007) A review of biodegradable thermoplastic starch polymers, pp 287–300
Hietala M, Mathew AP, Oksman K (2013) Bionanocomposites of thermoplastic starch and cellulose nanofibers manufactured using twin-screw extrusion. Eur Polymer J 49(4):950–956
Jabeen N, Majid I, Nayik GA (2015) Bioplastics and food packaging: a review. Cogent Food Agric 1(1):1117749
Juan M, Tirado G, Emilio OR, Juan CBR, Julio C, Tafolla A, Francisco HC, Juan BF (2024) Bionanocomposites based on thermoplastic starch for the food industry. Eng Princ Food Process Technol Prod Rization 1:29
Julinová M, Vaňharová L, Jurča M, Minařík A, Duchek P, Kavečková J, Skácelík P (2019) Effect of different fillers on the biodegradation rate of thermoplastic starch in water and soil environments. J Polym Environ 28(2):566–583
Jumaidin R, Sapuan SM, Jawaid M, Ishak MR, Sahari J (2016) Effect of seaweed on physical properties of thermoplastic sugar palm starch/agar composites. J Mech Eng Sci 10(3):2214–2225
Kampangkaew S, Thongpin C, Santawtee O (2014) The synthesis of cellulose nanofibers from SesbaniaJavanica for filler in thermoplastic starch. Energy Procedia 56:318–325
Khan B, Bilal Khan Niazi M, Samin G, Jahan Z (2017) Thermoplastic starch: a possible biodegradable food packaging material—a review. J Food Process Eng 40(3):1–17
Koch K, Gillgren T, Stading M, Andersson R (2010) Mechanical and structural properties of solution-cast high-amylose maize starch films. Int J Biol Macromol 46(1):13–19
Liu H, Xie F, Yu L, Chen L, Li L (2009) Thermal processing of starch-based polymers. Prog Polym Sci 34(12):1348–1368
Liu WC, Halley PJ, Gilbert RG (2010) Mechanism of degradation of starch, a highly branched polymer, during extrusion. Macromolecules 43(6):2855–2864
López OV, Castillo LA, Garcia MA, Villar MA, Barbosa SE (2015) Food packaging bags based on thermoplastic corn starch reinforced with talc nanoparticles. Food Hydrocoll 43:18–24
Lörcks J (1998) Properties and applications of compostable starch-based plastic material. Polym Degrad Stab 59(1–3):245–249
Lu Y, Weng L, Cao X (2005) Biocomposites of plasticized starch reinforced with cellulose crystallites from cottonseed linter. Macromol Biosci 5(11):1101–1107
Ma X, Yu J, Kennedy JF (2005) Studies on the properties of natural fibers-reinforced thermoplastic starch composites. Carbohyd Polym 62(1):19–24

Mahdieh Z, Bagheri R, Eslami M, Amiri M, Shokrgozar MA, Mehrjoo M (2016) Thermoplastic starch/ethylene vinyl alcohol/forsterite nanocomposite as a candidate material for bone tissue engineering. Mater Sci Eng C 69:301–310

Mano JF, Koniarova D, Reis RL (2003) Thermal properties of thermoplastic starch/synthetic polymer blends with potential biomedical applicability. J Mater Sci Mater Med 14(2):127–135

Melo CP, Grossmann MV, Yamashita F, Youssef EY, Dall'Antônia LH, Mali S (2011) Effect of manufacturing process and xanthan gum addition on the properties of cassava starch films. J Polym Environ 19(3):739

Mohammadi Nafchi A, Moradpour M, Saeidi M, Alias AK (2013) Thermoplastic starches: properties, challenges, and prospects. Starch-Stärke 65(1–2):61–72

Mondragón M, Hernández EM, Rivera-Armenta JL, Rodríguez-González FJ (2009) Injection molded thermoplastic starch/natural rubber/clay nanocomposites: morphology and mechanical properties. Carbohyd Polym 77(1):80–86

Mościcki L, Mitrus M, Wójtowicz A, Oniszczuk T, Rejak A, Janssen L (2012) Application of extrusion-cooking for processing of thermoplastic starch (TPS). Food Res Int 47(2):291–299

Müller CM, Laurindo JB, Yamashita F (2009) Effect of cellulose fibers addition on the mechanical properties and water vapor barrier of starch-based films. Food Hydrocoll 23(5):1328–1333

Munajad A, Subroto C (2018) Fourier transform infrared (FTIR) spectroscopy analysis of transformer paper in mineral oil-paper composite insulation under accelerated thermal aging. Energies 11(2):364

Muscat D, Adhikari B, Adhikari R, Chaudhary DS (2012) Comparative study of film forming behaviour of low and high amylose starches using glycerol and xylitol as plasticizers. J Food Eng 109(2):189–201

Myllymäki O, Eerikäinen T, Suortti T, Forssell P, Linko P, Poutanen K (1997) Depolymerization of barley starch during extrusion in water glycerol mixtures. LWT-Food Sci Technol 30(4):351–358

Nessi V, Falourd X, Maigret JE, Cahier K, D'orlando A, Descamps N, Lourdin D (2019) Cellulose nanocrystals-starch nanocomposites produced by extrusion: structure and behavior in physiological conditions. Carbohyd Polym 225:115123

Norrrahim MF, Janudin N, Asmal R, Mohd S, Jenol MA, Sharip NS, Nurazzi NM, Asyraf MRMI, Rushdan A (2024) Wheat thermoplastic starch composite films reinforced with nanocellulose. Phys Sci Rev 9(3):1509–1522

Orue A, Corcuera MA, Peña C, Eceiza A, Arbelaiz A (2016) Bionanocomposites based on thermoplastic starch and cellulose nanofibers. J Thermoplast Compos Mater 29(6):817–832

Perez-Pacheco E, Canto-Pinto JC, Moo-Huchin VM, Estrada-Mota IA, Estrada-León RJ, Chel-Guerrero L (2016) Thermoplastic starch (TPS)-cellulosic fibers composites: mechanical properties and water vapor barrier: a review. Compos Renew Sustain Mater. INTEACH, pp 85–105

Prachayawarakorn J, Pattanasin W (2016) Effect of pectin particles and cotton fibers on properties of thermoplastic cassava starch composites 38(2):129–136

Pushpadass HA, Hanna MA (2009) Age-induced changes in the microstructure and selected properties of extruded starch films plasticized with glycerol and stearic acid. Ind Eng Chem Res 48:8457–8463

Pushpadass HA, Kumar A, Jackson DS, Wehling RL, Dumais JJ, Hanna MA (2009) Macromolecular changes in extruded starch-films plasticized with glycerol, water and stearic acid. Starch-Stärke 61(5):256–266

Qing J, Zhipeng T, Huidong S, Yanfeng Z, Yucai S, Tingwei W (2022) Thermoplastic starch based blends as a highly renewable filament for fused deposition modeling 3D printing. Int J Biol Macromol 219:175–184

Raabe J, Fonseca ADS, Bufalino L, Ribeiro C, Martins MA, Marconcini JM, Tonoli GHD (2015) Biocomposite of cassava starch reinforced with cellulose pulp fibers modified with deposition of silica (SiO2) nanoparticles. J Nanomater 2015:1–9

Ren J, Dang KM, Pollet E, Avérous L (2018) Preparation and characterization of thermoplastic potato starch/halloysite nano-biocomposites: effect of plasticizer nature and nanoclay content. Polymers 10(8):808

Robert S, Ing K (2012) Thermoplastic starch, thermoplastic elastomers. RMIT University, Melbourne Australia, Applied Sciences

Sessini V, Arrieta MP, Raquez JM, Dubois P, Kenny JM, Peponi L (2019) Thermal and composting degradation of EVA/thermoplastic starch blends and their nanocomposites. Polym Degrad Stab 159:184–198

Sharip NS, Tengku YA, Tengku A, Farhana AU, Faiz N, Mohd N (2024) Potato thermoplastic starch nanocomposite films reinforced with nanocellulose. Phys Sci Rev 9(3):1479–1496

Shruthy R, Preetha R (2019) Cellulose nanoparticles from agro-industrial waste for the development of active packaging. Appl Surf Sci 484:1274–1281

Stavros XD, Orestis V, Zdenko Š, Hamed P, László L (2024) Thermoplastic starch processed under various manufacturing conditions: thermal and electrical properties. Biomacromol 25(9):5938–5948

Teixeira EDM, Pasquini D, Curvelo AA, Corradini E, Belgacem MN, Dufresne A (2009) Cassava bagasse cellulose nanofibrils reinforced thermoplastic cassava starch. Carbohyd Polym 78(3):422–431

Wang CZ, Li FY, Wang LM, Li JF, Guo AF, Zhang CW, Liu P (2015) Research on thermoplastic starch and different fiber reinforced biomass composites. RSC Adv 5(62):49824–49830

Willett JL, Millard MM, Jasberg BK (1997) Extrusion of waxy maize starch: melt rheology and molecular weight degradation of amylopectin. Polymer 38(24):5983–5989

Wunderlich B (2011) Do biopolymers behave the same as synthetic high polymers? J Therm Anal Calorim 106(1):81–84

Xie F, Luckman P, Milne J, McDonald L, Young C, Tu CY, Halley PJ (2014a) Thermoplastic starch. J Renew Mater 2(2):95–106

Xie F, Liu P, Yu L (2014b) Processing of plasticized starch-based materials: state of the art and perspectives. Starch Polym 257–289

Zain AHM, bin Ab Wahab MK, Ismail H (2019) Effect of calcium carbonate incorporation on the properties of low linear density polyethylene/thermoplastic starch blends. J Eng 15(2):97–108

Zélia A, Paulo B, Nuno M, Gonçalo F, Gonzalo OI, Idalina G, Sónia M, Paula F, Cláudia N (2024) Thermoplastic starch-based films loaded with biochar-ZnO particles for active food packaging. J Food Eng 361

Zhang Y, Han JH (2010) Crystallization of high-amylose starch by the addition of plasticizers at low and intermediate concentrations. J Food Sci 75:8–16

Zhang Y, Thompson M, Liu Q (2011) The effect of pea fiber and potato pulp on thermal property, surface tension, and hydrophilicity of starch-based bioplastics. Carbohyd Polym 86:700–707

Zhang CW, Nair SS, Chen H, Yan N, Farnood R, Li FY (2020) Thermally stable, enhanced water barrier, high strength starch bio-composite reinforced with lignin containing cellulose nanofibrils. Carbohyd Polym 230:115626

Zhang Y, Rempel C, McLaren D (2014b) Thermoplastic starch. In: Innovations in food packaging. Academic Press, pp 391–412

Zhang Y, Rempel C, Liu Q (2014a) Thermoplastic starch processing and characteristics—a review. Crit Rev Food Sci Nutr 54(10):1353–1370

Zhou J, Ma Y, Ren L, Tong J, Liu Z, Xie L (2009) Preparation and characterization of surface crosslinked TPS/PVA blend films. Carbohyd Polym 76(4):632–638

Chapter 7
Preparation, Characterization, and Antibacterial Activity of Biodegradable Bacterial Cellulose Nanocomposites

Ramesh Shruthy and Radhakrishnan Preetha

Abstract Nowadays, the usage of bacterial cellulose in food, paper, electronics, and biomedical industries has gained widespread attention. The main advantage of bacterial cellulose is its biocompatibility and biodegradability. Unlike the plant cellulose, the bacterial cellulose is devoid of lignin and hemicellulose, and hence, it is considered as the purest form of cellulose. Bacterial cellulose has nanofibrillar structures with large surface areas and is non-toxic in nature. Mainly, bacterial cellulose-based nano-biocomposites use in the biomedical industry, including wound dressing, which is discussed in this chapter. The moldability, high water-holding capacity, versatility, high mechanical strength and low-cost are the main reason for its usage in different products. This chapter discusses the type of microorganisms producing bacterial cellulose, the application of bacterial cellulose, preparation of bacterial cellulose, preparation of bacterial cellulose nanocomposite, studies on antibacterial activities of bacterial cellulose nanocomposites and the applications of bacterial cellulose nanocomposites in detail.

Keywords Bacterial cellulose · Antibacterial activity · Nanocomposites · Applications of bacterial cellulose nanocomposite

7.1 Introduction

Nowadays, microorganism-based cellulose production has gained remarkable attention. Bacteria such as Acetobacter, Rhizobium, Salmonella, Pseudomonas, *Acetobacter*, *Alcaligenes*, *Sarcina*, and *ventriculi* can synthesize nanofibrils of bacterial cellulose. The most efficient or commonly used bacteria for nano-fibrillated bacterial cellulose production are *Acetobacter xylinum*, *Acetobacter hansenii*, and *Acetobacter*

R. Shruthy · R. Preetha (✉)
Department of Food Process Engineering, School of Bioengineering, Faculty of Engineering and Technology, SRM Institute of Science and Technology, Kattankulathur, Tamil Nadu, India
e-mail: preethar@srmist.edu.in

P. M. Visakh (ed.), *Engineering Applications of Polymer-Based Nanocomposites*, Advanced Structured Materials 237, https://doi.org/10.1007/978-3-031-93463-6_7

pasteurianus (Klechkovskaya et al. 2008). *Acetobacter xylinum* can produce a high cellulose level and influence by the concentration of nitrogen and carbon sources (Hasanin et al. 2023; Moosavi-Nasab and Yousefi 2011). The production of the nano-fibrillated cellulose by *Acetobacter xylinum* involved the conversion of the polymerized glucose units into β (1 → 4) glucan chains (Iguchi et al. 2000). The bacterial cellulose can grow in static culture conditions; it will also produce thick pellicles with unique properties such as improved mechanical properties, high crystallinity, purity, and a homogeneous thick gel sheet-like structure (Klemm et al. 2005). Even though plant-based cellulose has the properties mentioned above, they are not pure cellulose; it contains lignin and hemicellulose. Various chemical and physical treatments do the isolation of the cellulose Nanofibers or cellulose Nanoparticle. The chemical treatment includes alkaline treatment, bleaching, and acid hydrolysis. Similarly, for physical treatment, ultra-sonication and ultra-homogenization are reported. Then the isolated cellulose Nanoparticles/fibers have amorphous and crystalline regions. The bacterial cellulose is a pure cellulose source and is used to prepare the composites or nanocomposites (Wu and Liu 2012).

The advantages of bacterial cellulose are shown in Fig. 7.1 (Klechkovskaya et al. 2008; Shah et al. 2013). Nano-fibrillated bacterial cellulose has various applications in the food industry, textile industry, cosmetic industry, paper industry, and medical/ medicine (Hasanin et al. 2023). Due to its unique property, bacterial cellulose uses as a natural source for the production of different biomedical materials such as tissue implants, scaffolds and materials for drug delivery. The study of Wu and Liu (2012); Saxena and Brown 2012) reports that the bacterial cellulose-based nanostructure incorporated a collegian can be used for cell support and cell immobilization in medicine. This chapter discusses the type of microorganisms producing bacterial cellulose, application of bacterial cellulose, preparation of bacterial cellulose, preparation of bacterial cellulose nanocomposite, studies on antibacterial activities of bacterial cellulose nanocomposites and the applications of bacterial cellulose nanocomposites in detail.

7.2 Types of Microorganisms Producing Bacterial Cellulose

The microbes that were producing bacterial cellulose are reported from the species of Agrobacterium, Gluconacetobacter, and Sarcina (Donini et al. 2010). The bacterium *Gluconobacter* can able to convert the ethanol to acetic acid (Güzel 2024).

An early report mentioned that some fungi, seaweed, and non-pathogenic bacteria such as *K. xylinus* could also produce bacterial cellulose (Brown 1886). The important property of *K. xylinus* is that it can produce extracellular cellulose with a variety of thicknesses by monitoring oxygenation. The major advantages of bacterial cellulose are non-toxic, biocompatible, and biodegradable (Klemm et al. 2001). The bacterial cellulose is unique in its native state compare to plant cellulose, and direct utilization is also possible. The plant cellulose is embedded in lignin, hemicellulose, pectin,

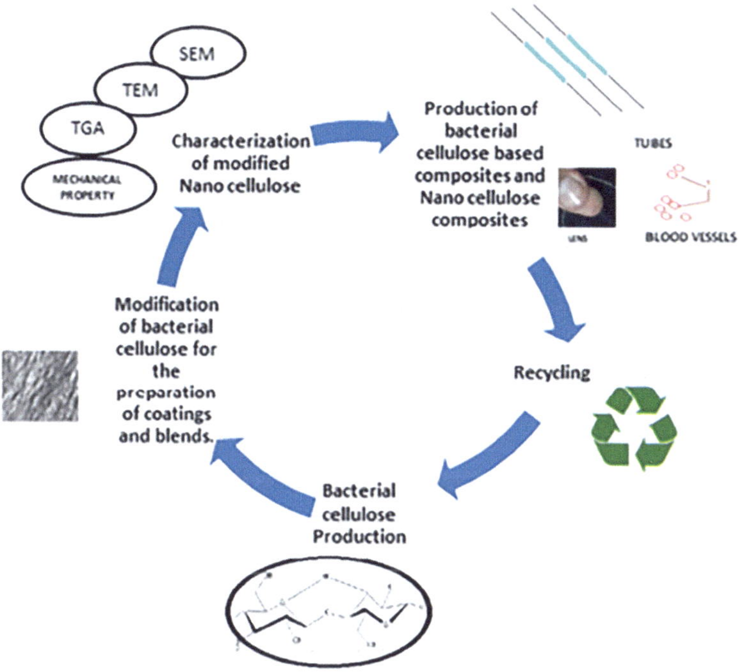

Fig. 7.1 Sustainable production of bacterial cellulose (Wu and Liu 2012)

and other related lignocellulosic matrices (Rahman and Netravali 2016). *Komagataeibacter xylinus* (*K. xylinus*) can produce two forms of bacterial cellulose, such as cellulose-1 and cellulose-2. Cellulose-1 is in the form of a ribbon-like structure. At the same time, cellulose-2 is a thermodynamically stable and amorphous polymer. Other bacterial cellulose producing bacteria in the genus *Komagataeibacter* are *Komagataeibacter kombucha/Komagataeibacter intermedium*. The genera such as *Achromobacter*, *Agrobacterium*, *Alcaligenes*, *Sarcina Rhizobium*, *Myxedema*, *Pseudomonas*, *Enterobacter*, *Salmonella*, and *Aerobacter*, are also reported for cellulose production.

7.3 The Large Scale Production of Bacterial Cellulose

Cellulose production by fermentation method involves several important factors, such as environmental conditions (temperature and humidity), growth medium, and byproducts (Ramana et al. 2000). The growth medium should contain a carbon source, nitrogen source, and other micronutrients. The carbon sources which are commonly used are sucrose and glucose. Other carbon sources, such as glycerol, maltose, and fructose, also have been reported (Chen et al. 2023). Nevertheless, the drawback while using glucose is the production of gluconic acid as a byproduct, which leads to a decrease in pH of the culture medium and results in the decreased production of bacterial cellulose (Ramana et al. 2000). Natural sources such as sugarcane molasses, juices of vegetables, and fruits containing a good carbon source are reported as culture media for bacterial cellulose production (Premjet et al. 2007; Molina-Ramírez et al. 2018). The reactor-based production of bacterial cellulose is not a commonly practicing procedure due to the lack of feasible technology for large-scale cellulose synthesis from bacterial. Similarly, the formation of the negative cellulose mutants in the reactor is also reported as an issue in bacterial cellulose production. The common reactors preferred are static and agitated bacterial cellulose culture, rotary biofilm contactor, spin filler attached bioreactors, rotary disk reactors, and silicon membrane-based reactors (Kim et al. 2007).

7.4 Characterization Study of Bacterial Cellulose and Cellulose-Based Composites

7.4.1 Morphological Studies

Bacterial cellulose and its composites are having improved mechanical, thermal, and optical properties (Tang et al. 2012). The identification of the morphological surface pattern of the bacterial cellulose can be determined by Scanning Electron Microscope (SEM). SEM is the most commonly used method for the morphological study for bacterial cellulose (Keshk 2011). SEM of a large depth field is used for the three-dimensional image of bacterial cellulose as well as bacterial cellulose Nanocomposites (Hong and Yang 2013). Sample preparation of the SEM analysis is a crucial step; the sample should not have water content. The normal air-drying process may change the structure of the bacterial cellulose (Tronser et al. 2018). For SEM analysis, the most commonly used drying methods are freeze-drying and critical-point drying (Helenius et al. 2006). It was reported that there was no visible difference between the critical-point drying and freeze-drying. After proper drying of the bacterial cellulose/bacterial cellulose Nanocomposite, it should stick over the Scanning Electron Microscope sample holder and followed by gold or platinum sputtering will take place followed by morphological analysis can be able to take place (Lina et al. 2016). Bacterial Nanocellulose morphological aspect study analysis and visibility of the clear structure of the bacterial cellulose pellicles are possible with polarized optical microscopy (Bodin et al. 2007). The Nano-sized width of the bacterial cellulose is identified by Transmission Electron Microscope (Picheth et al. 2017).

7.4.2 Mechanical and Physical Properties

Mechanical properties of the plant-derived cellulose are different from that of Bacterial cellulose. According to the study of Czaja et al. (2004), the major properties of bacterial cellulose are improved crystallinity index of 84–89%. Bacterial cellulose lack impurities compared to plant-based cellulose; hence, bacterial cellulose undertakes only a simple purification process (Sani and Dahman 2009). The mechanical properties and physical properties of bacterial cellulose will strongly be based upon the processing and manufacturing conditions. It was described that the dried bacterial cellulose is having a tensile strength of 240 MPa, elongation at break is 3%, and Young's modulus of around 10 GPa (Klemm et al. 2005). Similarly, another study claims that the dried bacterial cellulose sample has improved elongation at break of 10% (Fernandes et al. 2013). The above study infers that the production/manufacturing process of the bacterial cellulose and the nutritional source used for bacterial cellulose growth can improve the mechanical properties. A list of strains of microbes and its sources for bacterial cellulose production is shown in Table 7.1.

Table 7.1 List of strains of microbes and its sources for the production of bacterial cellulose

Species	Microorganism	Source of carbon	Supplement source	Refs.
BRC-5, E-25	*Acetobacter xylinum*	Glucose	Oxygen, ethanol	Czaja et al. (2006)
PJK (KCTC 10, 505 BP)	*G. hansenii*	Glucose	Oxygen, ethanol	Bae et al. (2004)
V6 and A9	*Acetobacter*	Glucose	Ethanol	Millon et al. (2008)
IFO 13773	*Acetobacter xylinus*	Glucose	Lignosulfonate	Bae and Shoda (2004)
NUST 4.1	*Acetobacter xylinus*	Glucose	Sodium alginate	Park et al. (2004)
K3	*G. xylinus*	Mannitol	Green tea	Touzel et al. (2003)
BRP2001	*Acetobacter xylinum*	Molasses and fructose	Oxygen	Kaewnoppara et al. (2008)
Sucrofermentans BRP2001	*Acetobacter species*	Fructose	Oxygen	Mondel et al. (2013)
IFO 13773	*G. xylinus*	Sugarcane molasses	–	Svensson et al. (2005)
RKY-5	*Gluconacetobacter*	Glycerol	–	Andrade et al. (2013)
St-60-12	*Gluconacetobacter*	Sucrose	–	George et al. (2005)
JCM116	*Lactobacillus mali*	Sucrose	–	Seto et al. (2006)

The bacterial cellulose of 98% moisture content was observed that it has an elongation at break of 21%, 380 kPa (tensile strength) and also has 2900 g m^{-1} water vapor transmission rate (Fernandes et al. 2017). By analyzing the morphological study of the hydrated bacterial cellulose, it was identified that it has a specific pore volume of 0.2 cm^3 g^{-1}, a specific surface area of 60 m^2 g^{-1}, and a 13 nm average pore diameter (Qiao et al. 2015). The above said property power to use for the production of the wound dressing in a biomedical application. The production/manufacturing of the bacterial cellulose-based wound dressing with post-synthetic modification and its advantages are shown in Fig. 7.2. It was reported that the aloe vera incorporated bacterial cellulose is also used for the wound dressing (Saibuatong and Phisalaphong 2010). It increases the water vapor permeability and water-holding capacity 1.5-fold than non-modified bacterial cellulose material. The other biomedical application includes Bacterial cellulose base skin moisturizer, burn to dress, and so on. The bacterial cellulose-based composites of 98.0% water content are not suitable for molding, whereas 98.5% of water content with aloe vera gel solution exhibits enhanced mechanical, improved water vapor transmission and hydrophilicity. The improved tensile strength and elasticity of the bacterial cellulose are used in the biomedical and food packaging, textile, and leather industries.

7.5 Cellulose Structure and Property Relationship

The naturally occurring cellulose can classify into two, bacterial cellulose and plant cellulose. Plant cellulose is mesh-like tough cellulose fibers, which can provide the plant with a basic plant structure (Sani and Dahman 2009). However, bacterial cellulose does not contain hemicellulose and lignin, which will be in the purest form of cellulose. The arrangement/structure of the bacterial cellulose is given in Fig. 7.3. The bacterial cellulose has good tensile strength, better crystallinity, good water-holding capacity, and high mobility/good meshwork formation during development (Iguchi et al. 2000).

Bacterial cellulose can be produced/molded-in sheet-shaped and having significant mechanical properties. Bacterial cellulose is macro-structured fibrils, which stacks in the form of bundles, then this makes it to ribbon form, and it is known as pellicle (Steinbuhel 2001). Oxygen is an essential factor for the growth of cellulose-producing microorganisms (Iguchi et al. 2000). The formation of the bacterial cellulose (BC) occurs on the topmost surface of the culture broth. The shape and size of the bacterial cellulose pellicle depend upon the surface area providing for the growth (Almihyawi et al. 2024).

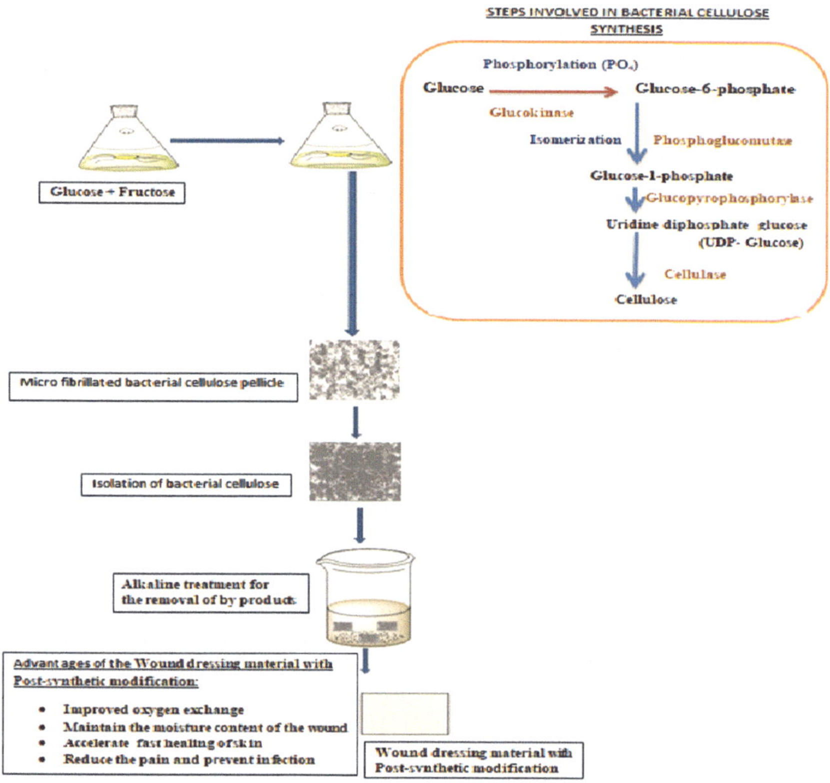

Fig. 7.2 Schematic representation of the bacterial cellulose based wound dressing with post-synthetic modification and its advantages (Qiao et al. 2015)

7.6 Biodegradation of the Bacterial Cellulose and Cellulose-Based Composites

Nowadays, the usage of plastic is banned because of environmental pollution due to prolonged decomposition time. In this context, the researchers are focusing on biodegradable plastics. The study of bacterial cellulose—poly(3-hydroxybutyrate) (PHB) has biodegradable nature by soil degradation method. The degradation study results in the complete degradation of the bacterial cellulose-based composite with 90 days. The calcium phosphate incorporated bacterial cellulose-polyvinyl pyrrolidone based hydrogel was studied by hydrolytic degradation method. The freeze-dried samples were placed inside the 25 mL of Ringer's solution of pH 7.4 and kept for the degradation study at 37 °C for four weeks. By considering the above studies, it confirms that the bacterial cellulose-based composites have improved biodegradability, and those are safe for the environment.

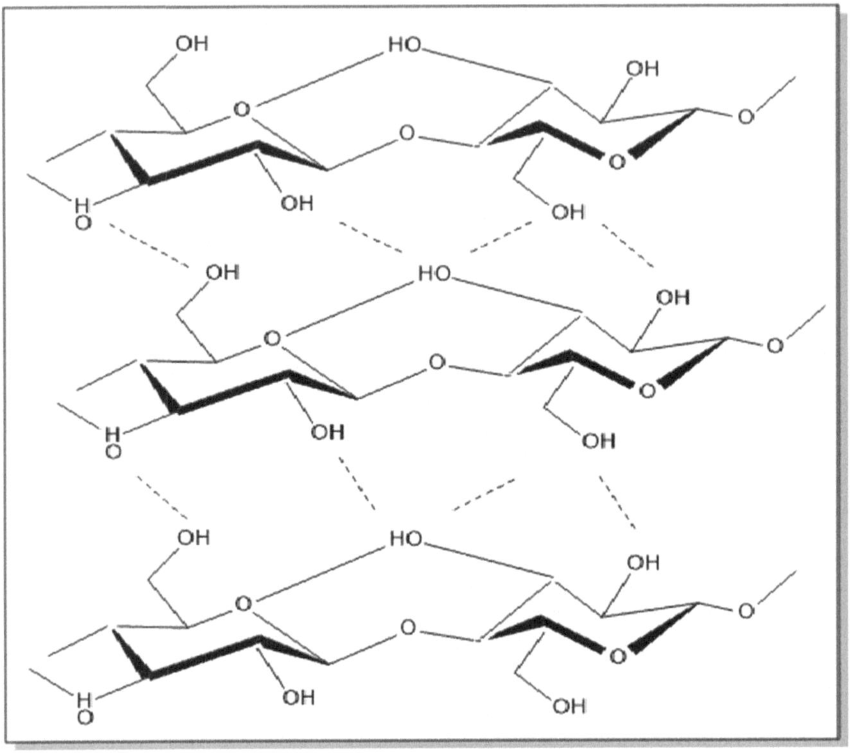

Fig. 7.3 The structure of bacterial cellulose

7.6.1 Preparation of Bacterial Cellulose Nanocomposites

The bacterial cellulose nanocomposites are mainly prepared to overcome the limitation of using bacterial cellulose alone. The bacterial cellulose does not have stress-bearing property, optical transparency, and antibacterial property (Santos et al. 2014). The bacterial cellulose with nano-based reinforcement can be able to improve the above-said limitations. The bacterial cellulose-based nanocomposites are synthesized by two methods they are ex-situ and in-situ method. In the case of in situ based synthesis, the nanomaterial is added along with the bacterial cellulose/polymer (Qiao et al. 2015). However, in the ex situ method, bacterial cellulose is impregnated with reinforcement materials (Ul-Islam et al. 2012). The nanomaterials of metal and metal oxides of Ag, Au, TiO_2, and carbonaceous Nanomaterials (carbon nanotubes, graphite) are commonly used to synthesize the bacterial cellulose Nanocomposites (Kim et al. 2013).

7.6.2 Studies on Antibacterial Activities of Bacterial Cellulose Nanocomposites

The bacterial cellulose Nanocomposites are mainly used in the medical, paper, textile, leather, and food industry, due to its sustainability, structure, and biocompatibility (Jebali et al. 2013). Moreover, for the medical industry, bacterial cellulose-based composites are used (Falcão et al. 2008). The bacterial cellulose Nanocomposites are capable of improving the antibacterial property, and it can be used for tissue engineering and drug delivery (Barud et al. 2011). In medicine, a bacterial cellulose-based wound dressing with antibacterial property is gaining attention nowadays (Vitta et al. 2010). The study reported that the bacterial cellulose with a porous network structured in dressing material could deliver medicine/antibiotics to the wound and create a barrier against external infection (Gao et al. 2014). The bacterial cellulose composites are synthesized by the addition of antimicrobial agents such as inorganic antimicrobials (ZnO, CuO, and Ag particles) and organic antimicrobials (enzymes) since bacterial cellulose are not having antimicrobial property (Luan et al. 2012). The inorganic antimicrobial agent, as silver Nanoparticle incorporated bacterial cellulose is having a major role in inhibiting bacteria. The Ag-bacterial cellulose Nanocomposites having the inhibition property on bacteria and the pathway for inhibition was also described in this study (Jebali et al. 2013; Shi et al. 2013). The attachment of Ag ions to the enzymes and the thiol group of protein is responsible for inhibiting bacteria. Hence it will affect the normal transportation and respiration mechanism of the bacterial cell. The silver ions (Ag) also attach to bacterial cells externally and internally (i.e., to the outer layer of the bacterial cell and inside bacterial cell membrane), thereby alter the bacterial cell wall function. Various studies prove the ability of the silver Nanoparticles to inhibit bacterial strains such as Escherichia coli, *Aspergillus niger*, fungal strains, and viruses (Rai et al. 2009).

7.6.3 Application of the Bacterial Cellulose and Cellulose Nanocomposites

The unique properties of bacterial cellulose have been used in diverse fields such as the food industry, the medical field, paper industry, and leather industry (Fig. 7.4).

The bacterial cellulose nanocomposites play a major role in the biomedical sector. Nowadays, bacterial cellulose (BC) is used as an alternative factor for the treatment and transplantation in tissue engineering. The studies show that bacterial cellulose has similar features to the human collagen (Markstedt et al. 2015). The major bacterial cellulose-based composites available in the market are medical dressing pads, cartilage, vascular grafts, dental implants, and artificial skin/bone. It is also used for the delivery of hormones, drugs, and proteins (Barbosa et al. 2016). The applications and functionality of the bacterial cellulose Nanocomposites in the biomedical field are given in Table 7.2.

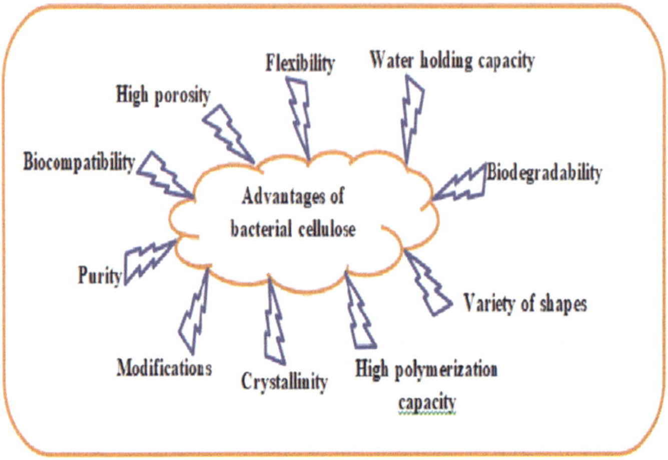

Fig. 7.4 Advantages of bacterial cellulose (Zhu et al. 2015)

7.6.4 Biomedical Application

In unhealed and absolute wounds, the bacterial cellulose-based Nanocomposite dressing bandages can reduce bacterial enzyme activity and prevent an anaerobic environment. Thereby reduce the infection and hence heal the chronic wound (Raquel et al. 2019; Wippermann et al. 2009). The first and foremost bacterial cellulose-based wound dressing product available on the market was Bio fill (Fibrocel Produtos Biotechnologies LTDA (Fibrocel), Brazil). The main properties of the biofill are its thin structure and 85% water content. Hence, it can be used to treat ulcers and burns as a substitute for skin/skin grafts. The main advantage of Bio fills it can fast relief pain and infection (Fontana et al. 1990). Later on, similar products for wound treatments in humans and veterinary were introduced in the market, for example, Xcell, Nano-derm, and cellulose (Blackburn 2005; Muangman et al. 2011). After introducing the bacterial cellulose-based wound dressing products, due to its good mechanical property, porosity, foldability, and compatibility, it started to be used to transplant the blood vessels with less than 5 mm (Lee and Park 2017; Innala et al. 2014). Nowadays, bacterial cellulose plays a major role in the cosmetic sector also. The definition of cosmetics is the product for the human body for beautifying, cleansing, and creating attractiveness without altering the body (Hasan et al. 2012). The bacterial cellulose has a role in the cosmetic formulation of oil-in-water emulsion since it does not cause any skin allergies. The allergy repellent nature of the bacterial cellulose is due to its hydration property, leading to various cosmetic creams. Similarly, bacterial cellulose has been used to prepare artificial nails, talcum powders, and so on. The invention of cosmetics for cleansing will provide good lather, easy rinse-off, and does not make the unpleasant slimy feel discussed by Gama et al. (2013).

In drug delivery, the USFDA itself approved that the bacterial cellulose-based composites can use without any side effects (Amin et al. 2014). The study of Perez

Table 7.2 Bio-medical applications and functions of bacterial cellulose nanocomposites

Application	Properties	Refs.
Medical application		
(a) Bacterial cellulose incorporated ZnO/TiO_2/Ag nanoparticle based nanocomposites for wound dressing	• Antimicrobial activity against *E-Coli, Staphylococcus aureus, Pseudomonas aeruginosa, Citrobacter freundii*	Khalid et al. (2016) Yang et al. (2012)
(b) Bacterial cellulose incorporated poly-L-lysine/tempo/silymarin-zein nanoparticle based nanoparticles based wound dressing	• 65% healing ability within 15 days by compare with initial day of dressing	Tsai et al. (2018) Stumpf et al. (2013)
(c) Artificial cornea E.g.: bacterial cellulose–PVA hyaluronic acid composites	• Increased tensile strength • Nontoxic • Tissue replacement and wound healing • Increased light transmittance	Hobzova et al. (2018) Lv et al. (2016)
(d) Blood vessels Bacterial cellulose–ion-oxide nanoparticle composites	• The values for blood vessels are based on Young's modulus	Bodin et al. (2010) Arias et al. (2016)
(e) Bacterial cellulose–chitosan nanocomposites	• Used as nano-films	Fernandes et al. (2009)
(f) Silver nanoparticle–bacterial cellulose composite	• Antimicrobial property	Zhang et al. (2013)
(g) Hydroxyapatite–bacterial cellulose composites	• Major component and essential ingredient for teeth and bone (maxilla facial)	Tazi et al. (2012)
(h) Bacterial cellulose–paraffin composites	• Enhanced mechanical property • Used for skin and tissue regeneration • Development of artificial organs • Used as skin substitutes	Meng et al. (2009) Kim et al. (2014)

et al. (2013) identified that bacterial cellulose incorporated with chloro-aluminium/pthalocyanine and photosensitizers could treat skin cancer. In Wu and Lia (2008) experiments, it was reported that the bacterial cellulose was used as an immobilizing material for glucoamylase. Immobilization of the glucoamylase enhanced stability at low pH and different temperatures. The bacterial cellulose has been used for bone grafts and bone-tissue engineering. Bone is made up of organic materials such as collagen/noncollagenous proteins and inorganic calcium hydroxyapatite. In a study, calcium chloride ($CaCl_2$) and sodium carbonate (Na_2CO_3) were added to the bacterial culture medium for promoting calcium carbonate deposition in bacterial cellulose. Calcium carbonate crystals incorporated bacterial cellulose enhanced the biocompatibility for applications in bone and tissue engineering (Stoica-Guzun et al. 2012). The ability of the bacterial cellulose to mimic collagen and Nanofiber has been proven for Ca–P deposition by mineralization. As the Ca–P minerals structure, the obtained platelet is similar to calcium-deficient hydroxyapatite, which has similar properties to hydroxyapatite present in real bone (Gao et al. 2010).

Similarly, in the study of Grande et al. (2009), it was reported that the bacterial cellulose was incorporated with hydroxyapatite Nanocomposite (hydroxyapatite and Carboxy Methyl Cellulose (CMC) was added to bacterial cellulose culture medium) with enhanced osteoconductivity and bio-compatibility. Bacterial cellulose is also reported for cardiovascular applications. There are blood vessels in the diameter range of 6–10 mm manufactured Dacron®. Nevertheless, for coronary, femoral arteries, and carotid artery, the need for small diameter sized blood vessels are preferred and developed bacterial cellulose-based blood vessels with a diameter less than 6 mm. Later, blood vessels of diameter 1 mm and thickness of 0.7 mm were also developed with bacterial cellulose and fibrin. The composite of glutaraldehyde treated bacterial cellulose—fibrin shows improved mechanical property than the blood vessels made from other materials (Saxena and Brown 2012; Klemm et al. 2001). The carotid artery implant study using bacterial cellulose-based blood vessels (4–6 mm) in the rat was successful, and within four weeks, the implanted blood vessel was covered with the connective tissues. This proves that the bacterial cellulose can be used as an alternative for the blood vessel (Klemm et al. 2001).

7.6.5 Food Industry Application

In food industry, the bacterial cellulose (BC) is used as a raw material for many dishes such as sweet candy, beverage, and desserts. Also, it is having the role of a thickener and stabilizing agent. As a thickener, it can maintain the consistency of the food items (Rajput et al. 2024). In Japan, the beverage known as Kombucha is prepared by the fermentation process, and bacterial cellulose is one of the additives used for the production process (Premjet et al. 2007). Kombucha tea is prepared by immersing/soaking tea leaves along with sugar in boiling water. The mixture of Kombucha and sugar is used to prepare the bacterial cellulose by the fermentation process. The SCOBY (symbiotic culture of bacteria and yeast) uses as a mother or biofilm for the bacterial cellulose synthesis. The biofilm will be placed inside the container filled with Kombucha tea-sugar mixture, and fermentation will occur. During fermentation, the mixture will produce sour and sweet flavor, alcohol, and acetic acid. Then the final product will be the bacterial cellulose (Iguchi et al. 2000). Bacterial cellulose is also used to produce innovative food products in South-East Asian countries, a good source of dietary fiber, and serve as a gelatinous cube known as nata-de-coco and nata-de-pina. The production flow charts for nata-de-coco and nata-de-pina preparations are shown in Fig. 7.5. The main properties of nata-de-coco are chewiness, smooth/soft textural properties, low fat, and low calories (Mesgari et al. 2024). Static culture of *Acetobacter xylinum* was reported to prepare nata-de-coco using coconut water as a carbon source (Jaganath et al. 2008; Khan et al. 2015).

Mesomaya et al. (2006) reported that the mix of nata-de-coco and cereal having high fiber would help reduce the lipid level of the consumer (Montealegre et al. 2012).

In the food sector, bacterial cellulose (BC) is mainly used for food packaging. The bacterial cellulose-based composites have properties such as enhanced mechanical/thermal/antibacterial property and biodegradability (Tang et al. 2012). In addition to the food industry, in electronics also the bacterial cellulose-based Nanocomposites play important roles. The applications and functionality of the bacterial cellulose Nanocomposites in the food and electronics industry are given in Table 7.3.

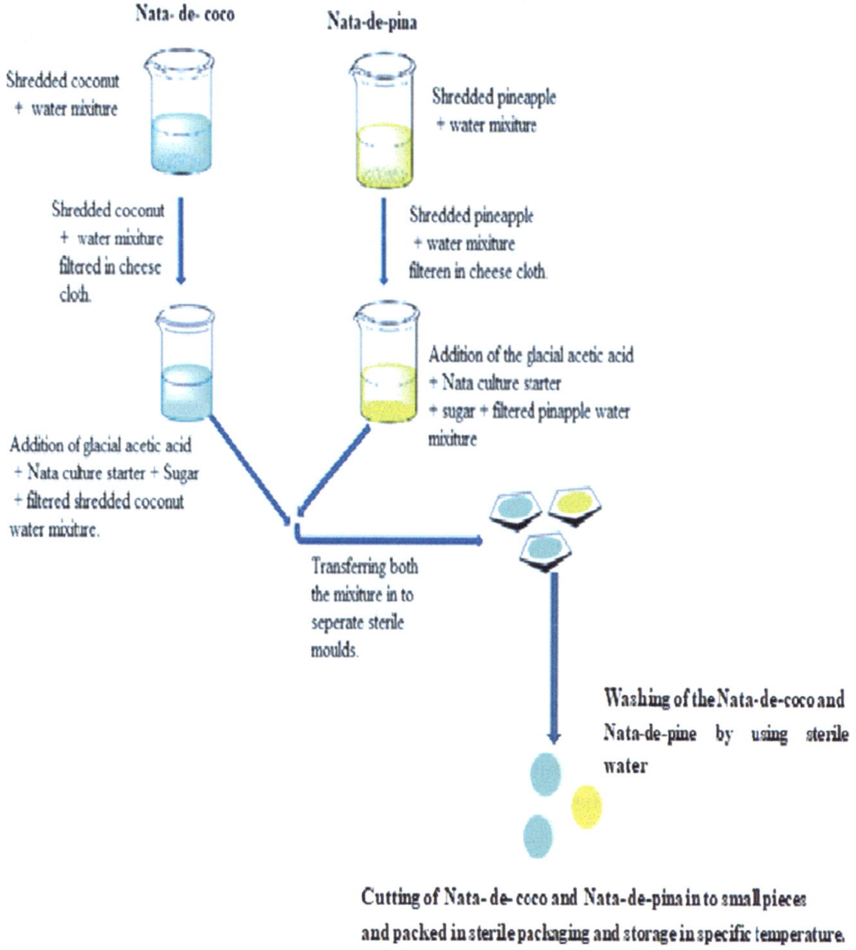

Fig. 7.5 Production flow chart of nata-de-coco and nata-de-pina (Khan et al. 2015; Mesomya et al. 2006)

Table 7.3 Applications and functions of bacterial cellulose nanocomposites in food and electronics industry

Application	Properties	Refs.
Food industry applications		
(a) Sausage packaging E.g.: bacterial cellulose-methyl cellulose composite	• Biodegradable	Kim et al. (2011)
(b) Whipping cream E.g.: hydroxy propyl-methyl cellulose	• To enhance the texture	Zhao et al. (2015) Olivera et al. (2016)
(c) Egg and vitamin-C preservative E.g.: methyl cellulose	• To improve shelf life and bioavailability	Nguyen et al. (2008) Zhu et al. (2015)
(d) Meat preservation E.g.: bacterial cellulose-nisin	• To improve antimicrobial property	Nguyen et al. (2008)
(e) Bacterial cellulose-chitosan nanocomposites	• Used as packaging material	Fernandes et al. (2009)
Electronics application		
(a) poly-urethane bacterial cellulose composites	• Used as a film substrate for light emitting diode	Ummartyotin et al. (2012) Gadim et al. (2014)
(b) Graphite nanoparticle—bacterial cellulose composites	• Outstanding mechanical, thermal and electrical properties. Used in batteries, sensors, supercapacitors	Ruhs et al. (2018)

7.6.6 Paper Industry

In paper industry, bacterial cellulose is used as a material that can improve the strength of the paper by coating on paper during processing and as a thickening agent for paper (Mande et al. 2012). Paper is a predominant product for every industry as well as for daily lives Paper can be made from various cellulose materials, including bacterial cellulose, plant-based cellulose, and so on (Taokaew 2024). The bacterial cellulose-based paper will improve the gloss, smoothness, and mechanical properties. Nowadays, various studies are undergone based on bacterial cellulose-based paper. A novel paper of bacterial cellulose-based electrically conductive carbon paper was developed. Furthermore, the bio-cellulose based magnetic paper with flexibility was also developed by Barud et al. (2015). The commercially available papers are more expensive for the ultrafiltration process so that the bacterial cellulose based nano papers for the ultrafiltration process was developed by Mautner et al. (2015).

7.6.7 Application of Bacterial Cellulose in Other Industries

In the cement manufacturing process, bacterial cellulose is used as an organic compound to prepare the hybrid composites. The bacterial nanocellulose incorporated cement was shown better hardening property than the conventional cement in a previous report (Busuioc et al. 2016). Bacterial cellulose is used as an electrically conducting material. It was reported that the bacterial cellulose incorporated carbon Nanotubes had its porous structure while immersing in an aqueous solution (Vitta and Thiruvengadam 2012). Then the electric conductivity was observed with carbon Nanotubes infiltrates present in the bacterial cellulose porous membrane. Bacterial cellulose—acrylic resin composite was developed, and it has been used to manufacture the flat-panel display of an organic LED (Nogi and Yano 2008). After considering the environmental protection, the researchers focus on the bacterial cellulose-based filters for wastewater treatment. It was reported that the filters made up of bacterial cellulose are having high porosity and elasticity. Hence, bacterial cellulose-based filters can be used for the mechanical treatment of the wastewater filtration process (Xiaobing 2017). Nowadays, in the case of biodegradable packaging, bacterial cellulose plays an important role. The packaging material made from bacterial cellulose can preserve and protect the food product (Tang et al. 2010). The property of the bacterial cellulose-based packaging material is increased water resistance, biodegradability, and are having a fine network of fibrils providing better mechanical property (Gao et al. 2014; Bayazidi et al. 2018). Bacterial cellulose has a vital role in the textile industry also reported the production of bacterial cellulose-based clothing. In the study, the Kombucha culture was used to grow bacterial cellulose and their ultimate goal as per their own words "Our ultimate goal is to grow a dress in a vat with a liquid." Their primary products from the bacterial cellulose are shoes, shirts, jackets, and body shapes (Montealegre et al. 2012; Jain and Gupta 2016). In the literature, it was reported that the Electronic manufacturer Sony Corporation, Japan introduced the World's first bacterial cellulose-based converter for headphones (Ivelovich and Leykin 2014). The main advantage of the headset as they are having the property of rigidity and roughness is the same as aluminum.

The various other users of bacterial cellulose in various industrials is given in Table 7.4

7.7 Conclusion

Bacterial cellulose preparation is an emerging area of research, and it has applications in the biomedical, electrical industry, food sector, and so on. The wide application potential of bacterial cellulose is due to its specific and unique structure. Incorporating the bacterial cellulose with the specific antibacterial agents, metals, and organic substances can enhance its property even in an unbelievable manner. There are many studies for the mass production of the bacterial cellulose and bacterial cellulose

Table 7.4 Use of bacterial cellulose in various industries (Gao et al. 2014; Zhao et al. 2015; Nogi and Yano 2008; Xiaobing 2017; Tang et al. 2010)

Sl. No.	Bacterial cellulose based materials/products	Applications
1	Biofill®, wound dressing, biosensors, skin grafting, blood vessels, heart valves	Bio-medical
2	Carbon paper, paper board with electric conducting materials, sign boards, synthetic paper	Paper
3	Thickener, nata-de-coco, nata-de-pina, Kampuchea tea, bread	Food
4	Jackets, shapers	Textile
5	Equipment's for sports	Sports
6	Moisturizer creams, health tonics, nail transplantation, nail polish, false tooth	Cosmetic
7	LED bulb holder, diaphragm of microphones, headset diaphragm	Electronics
8	Prevention of oil spills, As an absorbent for toxic pesticides and residues	Refinery and mining sewage
9	Body of car, planes and rockets	Automobile
10	Shoes, belt, wallets, bags	Leather

Nanocomposites preparation in a cost-effective way for further applications. In this chapter, it was explained about the preparation, characterization, and antibacterial activity of the bacterial cellulose Nanocomposites. The above information provides insight into various possible bacterial cellulose applications and helps discover novel bacterial cellulose composites with unique properties for an eco-friendly lifestyle.

References

Almihyawi RAH, Musazade E, Alhussany N, Zhang S, Chen H (2024) Production and characterization of bacterial cellulose by Rhizobium sp. isolated from bean root. Sci Rep 14(1):10848. https://doi.org/10.1038/s41598-024-61619-w

Amin MCIM, Ahmad N, Pandey M, Xin CJ (2014) Stimuli-responsive bacterial cellulose-g-poly(acrylic acid-co-acrylamide) hydrogels for oral controlled release drug delivery. Drug Dev Ind Pharm 40:1340–1349

Andrade FK, Alexandre N, Amorim I (2013) Studies on the biocompatibility of bacterial cellulose. J Bioact Compat Polym 28:97–112

Arias SL, Shetty AR, Senpan A, Echeverry-Rendón M, Reece LM, Allain JP (2016) Fabrication of a functionalized magnetic bacterial nanocellulose with iron oxide nanoparticles. J Vis Exp

Bae S, Sugano Y, Shoda M (2004) Improvement of bacterial cellulose production by addition of agar in a jar fermentor. J Biosci Bioeng 97:33–38

Barbosa A, Robles E, Ribeiro J, Lund R, Carreño N, Labidi J (2016) Cellulose nanocrystal membranes as excipients for drug delivery systems. Materials 9:1002; Khalid A, Khan R, Ul-Islam M, Khan T, Wahid F (2017) Bacterial cellulose-zinc oxide nanocomposites as a novel dressing system for burn wounds. Carbohydr Polym 164:214–221

Barud HS, Regiani T, Marques RFC, Lustri WR, Messaddeq Y, Ribeiro SJL (2011) Antimicrobial bacterial cellulose-silver nanoparticles composite membranes. J Nanomater 2011:1–8
Barud HS, Tercjak K, Gutierrez J, Viali WR, Nunes ES, Ribereiro SJL, Jafellici M, Nalin M, Marques RfC (2015) Biocellulose-based flexible magnetic paper. J Appl Phys 117(17):17B734
Bayazidi P, Almasi H, Asl AK (2018) Immobilization of lysozyme on bacterial cellulose nanofibers: characteristics, antimicrobial activity and morphological properties. Int J Biol Macromol 107:2544–2551
Blackburn R (2005) Biodegradable and sustainable fibres. Woodhead Publishing, Cambridge, UK
Bodin A, Bäckdahl H, Fink H, Gustafsson L, Risberg B, Gatenholm P (2007) Influence of cultivation conditions on mechanical and morphological properties of bacterial cellulose tubes. Biotechnol Bioeng 97:425–434
Bodin A, Bharadwaj S, Wu S, Gatenholm P, Atala A, Zhang Y (2010) Tissue-engineered conduit using urine-derived stem cells seeded bacterial cellulose polymer in urinary reconstruction and diversion. Biomaterials 31:8889–8901
Brown AJ (1886) XIX—The chemical action of pure cultivations of bacterium aceti. J Chem Soc Trans 49:172–187
Busuioc C, Stroescu M, Stoica-Guzun A, Voicu G, Jinga SI (2016) Fabrication of 3D calcium phosphates based scaffolds using bacterial cellulose as template. Ceram Int 42:15449–15458
Chen J, Hong F, Zheng H, Zheng L, Du B (2023) Using static culture method to increase the production of Acetobacter xylinum bacterial cellulose. J Nat Fibers 21(1). https://doi.org/10.1080/15440478.2023.2288286
Czaja W, Romanovicz D, Malcolm Brown R (2004) Structural investigations of microbial cellulose produced in stationary and agitated culture. Cellulose 11:403–411
Czaja W, Krystynowicz A, Bielecki S, Brownjr R (2006) Microbial cellulose the natural power to heal wounds. Biomaterials 27:145–151
de Oliveira Barud HG, da Silva RR, da Silva Barud H (2016) Multipurpose natural and renewable polymer in medical applications: bacterial cellulose. Carbohydr Polyr 153(33):406–420
Donini ÍAN, Salvi DTB, Fukumoto FK, Lustri WR, Barud HS, Marchetto R (2010) Biosynthesis and recent advances in production of bacterial cellulose. Eclética Quím 35:165–178
Falcão SC, Coelho ARDB, Evêncio Neto J (2008) Biomechanical evaluation of microbial cellulose (Zoogloea sp.) and expanded polytetrafluoroethylene membranes as implants in repair of produced abdominal wall defects in rats. Acta Cir Bras 23:184–191
Fernandes SCM, Oliveira L, Freire CSR, Silvestre AJD, Neto CP, Gandinia A, Desbrieres J (2009) Novel transparent nanocomposite films based on chitosan and bacterial cellulose. Green Chem 11:2023–2029
Fernandes SN, Almeida PL, Monge N, Aguirre LE, Reis D, de Oliveira CLP et al (2017) Mind the microgap in iridescent cellulose nanocrystal films. Adv Mater 29:1603560
Fernandes SN, Geng Y, Vignolini S, Glover BJ, Trindade AC, Canejo JP (2013) Structural color and iridescence in transparent sheared cellulosic films. Macromol Chem Phys 214:25–32
Fontana JD, De Souza AM, Fontana CK, Torriani IL, Moreschi JC, Gallotti BJ, Farah LFX (1990) Acetobacter cellulose pellicle as a temporary skin substitute. Appl Biochem Biotechnol 24–25:253–264
Gadim TDO, Figueiredo AGPR, Rosero-Navarro NC, Vilela C, Gamelas JAF, Barros-Timmons A, Neto CP, Silvestre AJD, Freire CSR, Figueiredo FML (2014) Nanostructured bacterial cellulose–poly(4-styrene sulfonic acid) composite membranes with high storage modulus and protonic conductivity. ACS Appl Mater Interfaces 6(10):7864–7875
Gama N, Gatenholm P, Klemm D (2013) Bacterial nanocellulose. CRC Press, Taylor and Francis Group, Boca Raton, FL
Gao C, Xiong GY, Luo HL, Ren KJ, Huang Y, Wan YZ (2010) Dynamic interaction between the growing CaP minerals and bacterial cellulose nanofibers during early biomineralization process. Cellulose 17:365–373

Gao C, Yan T, Du J, He F, Luo H, Wan Y (2014) Introduction of broad spectrum antibacterial properties to bacterial cellulose nanofibers via immobilising E-polylysine nanocoatings. Food Hydrocoll 36:204–211

George J, Ramana KV, Sabapathy SN, Jagannath JH, Bawa AS (2005) Characterization of chemically treated bacterial (Acetobacter xylinum) biopolymer: some thermo-mechanical properties. Int J Biol Macromol 37:189–194. https://doi.org/10.1016/j.ijbiomac.2005.10.007

Grande CJ, Torres FG, Gomez CM, Bañó MC (2009) Nanocomposites of bacterial cellulose/hydroxyapatite for biomedical applications. Acta Biomater 5:1605–1615

Güzel M (2024) Characterization of cellulose produced by bacteria isolated from different vinegars. Int J Biol Macromol 277(3):134436. https://doi.org/10.1016/j.ijbiomac.2024.134436

Hasan N, Biak DrA, Kamarudin S (2012) Application of bacterial cellulose (BC) in natural facial scrub. Int J Adv Sci, Eng, Inf Technol 2(4):1–4

Hasanin MS, Abdelraof M, Hashem AH, El Saied H (2023) Sustainable bacterial cellulose production by Achromobacter using mango peel waste. Microb Cell Factories 22:24. https://doi.org/10.1186/s12934-023-02031-3

Helenius G, Bäckdahl H, Bodin A, Nannmark U, Gatenholm P, Risberg B (2006) In vivo biocompatibility of bacterial cellulose. J Biomed Mater Res Part A 76A:431–438

Hobzova R, Hrib J, Sirc J, Karpushkin E, Michalek J, Janouskova O, Gatenholm P (2018) Embedding of bacterial cellulose nanofibers within PHEMA hydrogel matrices: tunable stiffness composites with potential for biomedical applications. J Nanomater 2018:1–11

Hong F, Yang G (2013) High-dietary-fiber biscuit containing bacterial cellulose and preparation method thereof. 7(9):756–766

Iguchi M, Yamanaka S, Budhiono A (2000) Bacterial cellulose—a masterpiece of nature's arts. J Mater Sci 35:261–270

Innala M, Riebe I, Kuzmenko V, Sundberg J, Gatenholm P, Hanse E, Johannesson S (2014) 3D Culturing and differentiation of SH-SY5Y neuroblastoma cells on bacterial nanocellulose scaffolds. Artif Cell Nanomed Biotechnol 42:302–308

Ivelovich M, Leykin A (2014) Nanocellulose and its application. J Sci Isr - Technol Adv 6(4):17–25

Jaganath A, Kalaiselvan A, Manjunatha SS, Raju PS, Bawa AS (2008) The effect of pH, source and ammonium sulphate concentration on the production of bacterial cellulose (nata-de-coco) by Acetobacter xylinum. World J Microboplogy Biotechnol 24(11):2593–2599

Jain P, Gupta C (2016) Textile recycling practices in India: a review. Int J Text Fash Technol 6:21–36

Jebali A, Hekmatimoghaddam S, Behzadi A, Rezapor I, Mohammadi BH, Jasemizad T, Sayadi M (2013) Antimicrobial activity of nanocellulose conjugated with allicin and lysozyme. Cellulose 20:2897–2907

Kaewnopparat S, Sansernluk K, Faroongsarng D (2008) Behavior of freezable bound water in the bacterial cellulose produced by Acetobacter xylinum: an approach using thermoporosimetry. AAPS PharmSciTech 9:701–707

Keshk S (2011) Scope on bacterial cellulose production. LAP LAMBERT Academic Publishing, Saarbrücken

Khan S, Ul-Islam M, Khattak WA, Ullah MW, Park JK (2015) Bacterial cellulose-titanium dioxide nanocomposites: nanostructural characteristics, antibacterial mechanism, and biocompatibility. Cellulose 22:565–579

Kim J, Vai Z, Lee HS, Choi GS, Lee DH (2011) Preparation and characterization of novel bacterial cellulose/chitosan composite for potential biomedical application. J Polym Res 18(4):739–744

Kim SH, Lee CM, Kafle K (2013) Characterization of crystalline cellulose in biomass: basic principles, applications, and limitations of XRD, NMR, IR, Raman, and SFG. Korean J Chem Eng 30:2127–2141

Kim H, Nam S, Jeong J, Lee S, Seo J, Han H, Kim Y (2014) Organic solar cells based on conjugated polymers: history and recent advances. Korean J Chem Eng 3(1):1095–1104

Kim JY, Kim JN, Wee YJ, Park DH, Ryes HW (2007) Bacterial cellulose production by cluconacetobacter sp. RKY5 in a rotary biofilm contractor. Appl Biochem Biotechnol 137–140(1–12):529–537

Klechkovskaya VV, Volkov VV, Shtykova EV, Arkharova NA, Baklagina YG, Khripunov AK, Smyslov RY, Borovikova LN, Tkachenko AA (2008) Network model of Acetobacter xylinum cellulose intercalated by drug nanoparticles. In: Nanomaterials for application in medicine and biology. Springer, New York, NY, USA, pp 165–177

Klemm D, Schumann D, Udhardt U, Marsch S (2001) Bacterial synthesized cellulose: artificial blood vessels for microsurgery. Prog Polym Sci 26:1561–1603

Klemm D, Heublein B, Fink HP, Bohn A (2005) Cellulose: fascinating biopolymer and sustainable raw material. Angew Chem - Int Ed 44(22):3358–3393

Lee SE, Park YS (2017) The role of bacterial cellulose in artificial blood vessels. Mol Cell Toxicol 13:257–261

Lina F, Chandra P, Adrianna M, Wankei W (2016) Bacterial cellulose production using a novel microbe. Front Bioeng Biotechnol 4

Luan J, Wu J, Zheng Y, Song W, Wang G, Guo J, Ding X (2012) Impregnation of silver sulfadiazine into bacterial cellulose for antimicrobial and biocompatible wound dressing. Biomed Mater 7:65006

Lv X, Yang J, Feng C, Li Z, Chen S, Xie M, Xu Y (2016) Bacterial cellulose-based biomimetic nanofibrous scaffold with muscle cells for hollow organ tissue engineering. ACS Biomater Sci Eng 2:19–29

Mande BM, Blok K, Patel MK (2012) Innovation in paper making: LCA of printing and writing paper from conventional and high yield pulp. Sci Total Environ 439(4):307–320

Markstedt K, Mantas A, Tournier I, Ávila HM, Hägg D, Gatenholm P (2015) 3D Bioprinting human chondrocytes with nanocellulose-alginate bioink for cartilage tissue engineering applications. Biomacromolecules 16:1489–1496

Mautner A, Lee KY, Tammelin T, Mathew AP, Nedoma AJ, Li K, Bismark A (2015) Cellulose nanopaper as tight aqueous ultra-filtration membranes. React Funct Polym 86(4):209–214

Meng N, Zhou NL, Zhang SQ, Shen J (2009) Synthesis and antimicrobial activities of polymer/montmorillonite–chlorhexidine acetate nanocomposite films. Appl Clay Sci 42:667–670

Mesgari M, Matin MM, Goharshadi EK, Mashreghi M (2024) Biogenesis of bacterial cellulose/xanthan/CeO_2NPs composite films for active food packaging. Int J Biol Macromol 273(2):133091. https://doi.org/10.1016/j.ijbiomac.2024.133091

Mesomya W, Pakpeankitvatana V, Komindr S, Leelahakul P, Cuptapun Y, Hengsawadi D, Tammarate P, Tangkanakul P (2006) Effects of health food from cereal and nata de coco on serum lipids in human. Songklanakarin J Sci Technol 28(1):23–28

Millon LE, Guhados G, Wan W (2008) Anisotropic polyvinyl alcohol bacterial cellulose nanocomposite for biomedical applications. J Biomed Mater Res Part B 86:444–452

Molina-Ramírez C, Enciso C, Torres-Taborda M, Zuluaga R, Gañán P, Rojas OJ, Castro C (2018) Effects of alternative energy sources on bacterial cellulose characteristics produced by Komagataeibacter medellinensis. Int J Biol Macromol 117:735–741

Mondal MIH (2013) Mechanism of structure formation of microbial cellulose during nascent stage. Cellulose 20:1073–1088

Montealegre CM, Dionisio ER, Sumera LV, Adolacion JRT, Leon RD (2012) Continuous bioethanol production using saccharomyces cerevisiae cells immobilized in nata de coco (biocellulose). IPCBEE (35):77–81

Moosavi-Nasab M, Yousefi M (2011) Biotechnological production of cellulose by *Gluconacetobacter xylinus* from agricultural waste. Iran J Biotechnol 9(2):94–101

Muangman P, Opasanon S, Suwanchot S, Thangthed O (2011) Efficiency of microbial cellulose dressing in partial-thickness burn wounds. J Am Coll Certif Wound Spec 3(1):16–19

Nguyen VT, Gidley MJ, Dykes GA (2008) Potential of a nisin-containing bacterial cellulose film to inhibit Listeria monocytogenes on processed meats. Food Microbiol 25:471–478

Nogi M, Yano H (2008) Transparent nanocomposites based on cellulose produced by bacteria offer potential innovation in the electronic device industry. Adv Mater 20(10):1849–1852

Park HO, Bang YB, Joung HJ, Kim BC, Kim HR (2004, October 26) Lactobacillus KCTC 0774BP and Acetobacter KCTC 0773BP for treatment or prevention of obesity and diabetes mellitus. U.S. Patent 6808703

Pérez CD, De'Nobili MD, Rizzo SA, Gerschenson LN, Descalzo AM, Rojas AM (2013) High methoxyl pectin–methyl cellulose films with antioxidant activity at a functional food interface. J Food Eng 116:162–169

Picheth GF, Pirich CL, Sierakowski MR, Woehl MA, Sakakibara CN, de Souza CF, de Freitas RA (2017) Bacterial cellulose in biomedical applications: a review. Int J Biol Macromol 104:97–106

Premjet S, Premjet D, Ohtani Y (2007) The effect of ingredients of sugar cane molasses on bacterial cellulose production by Acetobacter xylinum ATCC 10245. Sen-I-Gakkaishi 63(8):193–199

Qiao K, Zheng YD, Guo SL, Tan J, Chen XH, Li JQ (2015) Hydrophilic nanofiber of bacterial cellulose guided the changes in the micro-structure and mechanical properties of nf-BC/PVA composites hydrogels. Compos Sci Technol 118:47–54

Rahman MM, Netravali AN (2016) Aligned bacterial cellulose arrays as "Green" nanofibers for composite. ACS Macro Lett 5(9):1070–1074

Rai M, Yadav A, Gade A (2009) Silver nanoparticles as a new generation of antimicrobials. Biotechnol Adv 27:76–83

Rajput M, Pamecha D, Kumari P, Chaturvedi P, Sharma C, Mishra R, Chauhan S (2024) Bio-bleaching of ankara pulp with xylanase-producing bacterial consortium for sustainable hand-made paper production. Curr Res Microb Sci 7:100262. https://doi.org/10.1016/j.crmicr.2024.100262

Ramana KV, Singh L, Lokendra (2000) Effect of various carbon and nitrogen sources on cellulose synthesis by Acetobacter xylinum. World J Microbiol Biotechnol 16(3):245–248

Raquel P, Catarina RL, Pedro L, Almeida, Rita GS (2019) Bacterial cellulose: a versatile biopolymer for wound dressing applications. Microb Biotechnol 12(4):586–610

Rühs PA, Storz F, Gómez YAL, Haug M, Fischer P (2018) 3D Bacterial cellulose biofilms formed by foam templating. NPJ Biofilms Microbiomes 4:21

Saibuatong OA, Phisalaphong M (2010) Novo aloe vera-bacterial cellulose composite film from biosynthesis. Carbohyd Polym 79(2):455–460

Sani A, Dahman Y (2009) Improvements in the production of bacterial synthesized biocellulose nanofibres using different culture methods. J Chem Technol Biotechnol 85(2):151–164

Santos SM, Carbajo JM, Quintana E, Ibarra D, Gomez N, Ladero M, Eugenio ME, Villar JC (2024) Characterization of purified bacterial cellulose focused on its use on paper restoration. Carbohydr Polym. https://doi.org/10.1016/j.carbpol.2014.03.064

Saxena IM, Brown RM (2012) Biosynthesis of bacterial cellulose. In: Bacterial nano cellulose: a sophisticated multifunctional material. CRC Press, Boca Raton, FL, USA, pp 1–18

Seto A, Saito Y, Matsushige M, Kobayashi H, Sasaki Y, Tonouchi N, Tsuchida T, Yoshinaga F, Ueda K, Beppu T (2006) Effective cellulose production by a coculture of Gluconacetobacter xylinus and Lactobacillus mali. Appl Microbiol Biotechnol 73:915–921

Shah N, Ul-Islama M, Khattaka WA, Parka JK (2013) Overview of bacterial cellulose composites: a multipurpose advanced material. Carbohyd Polym 98:1585–1598

Shi QS, Feng J, Li WR, Zhou G, Chen AM, Ouyang YS, Chen YB (2013) Effect of different conditions on the average degree of polymerization of bacterial cellulose produced by Gluconacetobacter intermedius BC-41. Cellul Chem Technol 47(7–8):503–508

Steinbuhel (2001) Bacterial cellulose. In: Biopolymers. Wiley-VCH, Weinheim

Stoica-Guzun A, Stroescu M, Jinga S, Jipa I, Dobre T, Dobre L (2012) Ultrasound influence upon calcium carbonate precipitation on bacterial cellulose membranes. Ultrason Sonochemistry 19:909–915

Stumpf TR, Pértile RAN, Rambo CR, Porto LM (2013) Enriched glucose and dextrin mannitol-based media modulates fibroblast behavior on bacterial cellulose membranes. Mater Sci Eng C 33:4739–4745

Svensson A, Nicklasson E, Harrah T, Panilaitis B, Kaplan DL, Brittberg M, Gatenholm P (2005) Bacterial cellulose as a potential scaffold for tissue engineering of cartilage. Biomaterials 26:419–431

Tang W, Jia S, Jia Y, Yang H (2010) The influence of fermentation conditions and post-treatment methods on porosity of bacterial cellulose membrane. World J Microbiol Biotechnol 26:125

Tang X, Kumar P, Alavi S, Sandeep K (2012) Recent advances in biopolymers and biopolymer-based nanocomposites for food packaging materials. Crit Rev Food Sci Nutr 52:426–442

Taokaew S (2024) Bacterial nanocellulose produced by cost-effective and sustainable methods and its applications: a review. Fermentation 10(6):316. https://doi.org/10.3390/fermentation10060316

Tazi N, Zhang Z, Messaddeq Y, Almeida-Lopes L, Zanardi LM, Levinson D, Rouabhia M (2012) Hydroxyapatite bioactivated bacterial cellulose promotes osteoblast growth and the formation of bone nodules. AMB Express 2:61

Touzel J-P, Chabbert B, Monties B, Debeire P, Cathala B (2003) Synthesis and characterization of dehydrogenation polymers in Gluconacetobacter xylinus cellulose and cellulose/pectin composite. J Agric Food Chem 51:981–986

Tronser T, Laromaine A, Roig A, Levkin PA (2018) Bacterial cellulose promotes long-term stemness of mESC. ACS Appl Mater Interfaces 10:16260–16269

Tsai Y-H, Yang Y-N, Ho Y-C, Tsai M-L, Mi F-L (2018) Drug release and antioxidant/antibacterial activities of silymarin-zein nanoparticle/bacterial cellulose nanofiber composite films. Carbohydr Polym 180:286–296

Ul-Islam M, Khan T, Park JK (2012) Nanoreinforced bacterial cellulose–montmorillonite composites for biomedical applications. Carbohydr Polym 89:1189–1197

Ummartyotin S, Juntarob J, Sainb M, Manuspiyaa H (2012) Development of transparent bacterial cellulose nanocomposite film as substrate for flexible organic light emitting diode (OLED) display. Ind Crops Prod 35:92–97

Vitta S, Thiruvengadam V (2012) Multifunctional bacterial cellulose and nanoparticles-embedded composites. Curr Sci 102(10):1398–1405

Vitta S, Drillon M, Derory A (2010) Magnetically responsive bacterial cellulose: synthesis and magnetic studies. J Appl Phys 108:53905

Wippermann J, Schumann D, Klemm D, Kosmehl H, Salehi-Gelani S, Ahlers T (2009) Preliminary results of small arterial substitute performed with a cylindrical biomaterial composed of bacteria cellulose. Eur J Vasc Endovasc Surg 37:592–596

Wu SC, Lia YK (2008) Application of bacterial cellulose pellets in enzyme immobilization. J Mol Catal B Enzym 54:103–108

Wu J-M, Liu R-H (2012) Thin stillage supplementation greatly enhances bacterial cellulose production by *Gluconacetobacter xylinus*. Carbohyd Polym 90(1):116–121

Xiaobing L (2017) Soy protein isolate bacterial cellulose composite membranes for high efficiency particulate air filtration. Compos Sci Technol 138:124–133

Yang G, Xie J, Deng Y, Bian Y, Hong F (2012) Hydrothermal synthesis of bacterial cellulose/AgNPs composite: a 'green' route for antibacterial application. Carbohydr Polym 87:2482–2487

Zhang Y, Nypelö T, Salas C, Arboleda J, Hoeger IC, Rojas OJ (2013) Cellulose nanofibrils. J Renew Mater 1:195–211

Zhao C, Li Z, Li T, Zhang Y, Bryant DA, Zhao J (2015) High-yield production of extracellular type-I cellulose by the cyanobacterium *Synechococcus* sp. PCC 7002. Cell Discov 1:15004

Zhu W, Li W, He Y, Duan T (2015) In-situ biopreparation of biocompatible bacterial cellulose/graphene oxide composites pellets. Appl Surf Sci 338:22–26

Chapter 8
Different Types of Biodegradation Analysis for Nanocomposites

R. Arun, R. Shruthy, and R. Preetha

Abstract Nanocomposites are produced to enhance the properties of a polymer matrix by adding suitable nanomaterials. Biodegradability is an important factor to be considered when a nanocomposite is produced in larger quantities for various applications. Lack of biodegradable property will result in the same accumulation, which can end up in serious environmental problems. Added nanoparticles also greatly influence the biodegradability of nanocomposites based on their organic or inorganic nature. Hence, various methods were developed to check the degradation rate of the nanocomposites and are discussed in detail in this chapter. Soil, water, and bacterial degradation studies are of more importance as it predicts the time of accumulation of nanocomposite under the soil and sea after their disposal. Other degradation studies described are based on the type of agents used for degradation, including enzymes, light radiation, IR radiation, mechanical methods, oxidation methods, chemical methods, and ultrasonic wave degradations. This chapter also describes various analysis methods for biodegradation used for the degradation rate analysis of bio nanocomposite, and the design suggested economical ways of degradation.

Keywords Nanocomposites · Biodegradability · Soil degradation · Enzymatic degradation · Photodegradation

8.1 Introduction on Biodegradation Analysis

Nanomaterials are materials that exhibit incredible thermal/mechanical/optical properties and design aspects. Nanocomposites are composed of nanomaterials as one component in the nanometer range. Novel nanocomposites explore wide business opportunities in different industrial sectors. Renewable resources or natural sources are mainly used in nanocomposites, leading to increasing biodegradability

R. Arun · R. Shruthy · R. Preetha (✉)
Department of Food Process Engineering, School of Bioengineering, Faculty of Engineering and Technology, SRM Institute of Science and Technology, Kattankulathur, Tamil Nadu, India
e-mail: preethar@srmist.edu.in

P. M. Visakh (ed.), *Engineering Applications of Polymer-Based Nanocomposites*, Advanced Structured Materials 237, https://doi.org/10.1007/978-3-031-93463-6_8

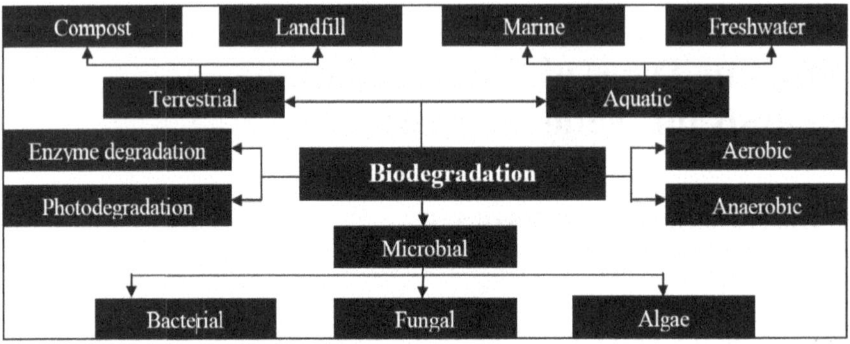

Fig. 8.1 Types of possible biodegradations for nanocomposites (Briassoulis and Dejean 2010; Narayn and Narayn 1999)

(Rutkowska et al. 1998). Natural sources for making the nanocomposites achieve desirable properties such as biodegradability, improved thermal, optical, and mechanical properties. Usage of organic nanoparticles (cellulose, chitosan, and starch) increases the biodegradability compared to inorganic nanoparticles (silver, titanium, and copper) in the nanocomposite. The biodegradable property can solve the environmental issues created by non-renewable sources and reduce the usage of synthetic polymers. There are different biodegradation types based upon the sources used to compose the Nanocomposites (EN 13432 2000) (Fig. 8.1). The degradation process can be achieved naturally by microbes, which are present in the environment, and the byproduct will be less harmful. The degradability analysis for the Nanocomposites is mainly performed by soil burial and compost methods (Kale et al. 2007). This chapter provides a better understanding of biodegradation studies such as water degradation, soil degradation, bacterial degradation, enzyme degradation, mechanical degradation, chemical degradation, degradation by IR radiation, ultrasonic degradation, and photo degradations. The biodegradation method can be chosen based on the nature and usage of composite and the possible place available for waste disposal.

8.2 Water Degradation Studies

Nowadays, it is a major concern that plastic and other synthetic polymers are floating over or deposited in the water bodies, and this leads to an imbalance in the water ecosystem. Many studies are undergoing to determine the water degradation condition in the anoxic sediment, deep seafloor, and supralittoral zone. The water degradation study is mainly conducting in marine habitats, and its classification is given in Fig. 8.2.

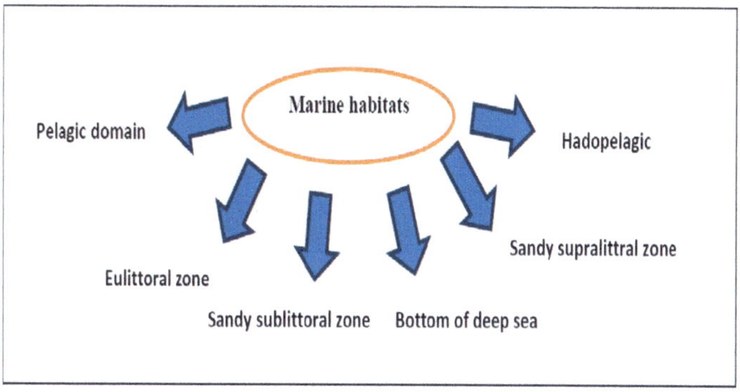

Fig. 8.2 Classification of degradation zones in the marine habitat (Tosin et al. 2012)

8.2.1 Pelagic Domain-Based Degradation Study

This is a laboratory-based test, which can carry for the determination of the condition of the composites/sample in a pelagic domain. Pelagic domain means it is the condition in which an aquarium is filled with saltwater. The test samples were packed in net-based pouches (the net pouches should be non-biodegradable). Then the packets of samples are immersed in the aquarium with an approximate length of 2 cm. Inside the aquarium, artificially created water circulation is one of the important factors, and it will help the degradation of the sample. The temperatures and salinity are the factors that will help for degradation; everyday checking of the salinity and adjustment is necessary throughout the study. For sampling study, the specimen should be a takeoff from the controlled conditions at regular intervals and will undergo physical analysis such as tensile strength and elongation at break (Rutkowska et al. 2002). There are various water-based degradation studies carried out for nanocomposites as well as LDPE (Low-Density Polyethylene), PE (Polyethylene) based packaging materials (Tsuji and Suzuyoshi 2002).

8.2.2 Eulittoral Zone-Based Biodegradation Study

This is a kind of degradation study carried out in the coastal region of the marine area. The main purpose of this study is to measure the physical degradation of the sample with the help of tides in the ocean/sea. The study is carried out at the borderline of the sea or ocean, which means the area should be constantly damp with the help of tides (Tosin et al. 2012; Briassoulis and Pikasi 2024).

The same study can also be carried out in laboratory conditions with the help of a reactor filled with sand and seawater (Rutkowska et al. 2002). The reactor should also be kept in a closed condition to avoid excessive drying out and vapor release. The sand and seawater should be collected from the littoral zone to contain an increased organic matter. After creating the preferred condition, the biodegradation study sample was placed in a framework of specific parameters and buried in the sandy bed. Then, the sample weights were taken at regular intervals of time (Weber et al. 2006).

8.2.3 Sandy Sublittoral Zone-Based Degradation Study

This is the study method based on the biological oxygen demand test (ISO 14851 1999). In this test, the powdered samples were introduced with the activated sludge of the sewage plant, and the determination of BOD will be determined. In the sandy sublittoral zone-based biodegradation study, the sediment and seawater were collected from the sandy sublittoral zone and kept inside the reactor, as mentioned before, eulittoral zone degradation study. But in this case, the study will be carryout at room temperature (Rutkowska et al. 2002; Liang et al. 2024).

The nanocomposite samples (specimens) were cut into pieces, and it has to keep inside the net-bags, which are non-biodegradable. Those bags, along with specimens, have to keep inside the reactor. After that, the sediments should be filled inside the reactor, and based on the wet weight of the sediment, seawater/synthetic seawater should also be added to the reactor in 1:5 ratios. Then, the theoretical oxygen demand inside the closed reactor will be determined by measuring the oxygen consumed by the sample at regular intervals (ISO 14851 1999). The glass-made reactors are used to determine the mineralization and disintegration of the samples (Takishita et al. 2006).

Microbes and fungi are helping in the degradation of the plastics, as well as Nanocomposites. The role of mycelial microbes for the degradation was reported in the research paper of Cosgrove et al. (2007). In the above case, composite (polyurethane) buried in the sand and fungal counts on the composite and sand was determined by plating on soil extract agar after serial dilution. Agar is also added with chloramphenicol (50 μg/mL) to inhibit the bacterial growth. The plates were incubated for 5–7 days at 250 °C. Total viable fungal counts can be driven from soil extract agar while colonies able to degrade can be counted from clearance zones in impranil agar. Thus the number of fungi capable of degradation (PU) can be calculated as a percentage of the total viable count. Different polymer degradation can be studied by taking respective agars instead of impranil.

8.3 Soil Degradation Studies

The soil burial method is an important method for disposing of the packaging material waste. The waste will degrade over time and reduces the waste mass. Bio-nanocomposite has the advantage of faster degradation than normal plastics, thereby reducing the accumulation frequency. It is thus important to study the soil biodegradability of the bio nanocomposite produced. Different methods are carried out so far, which can be chosen based upon the material produced and equipment

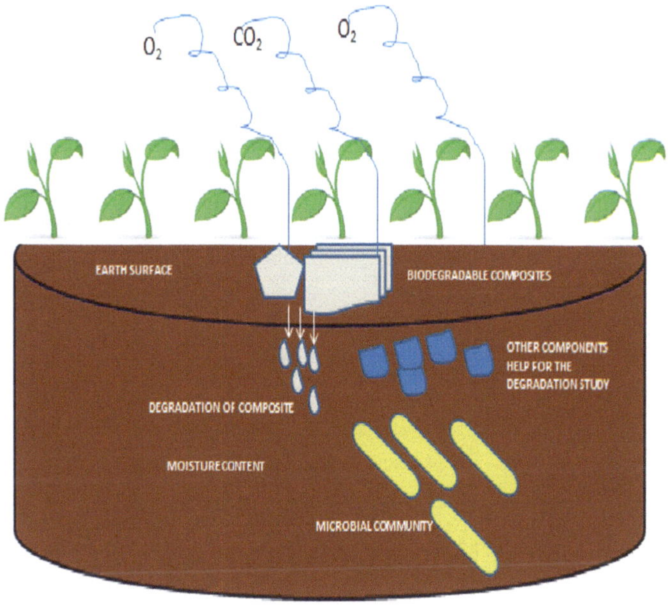

Fig. 8.3 Schematic representation of the soil degradation of composites

available (Raghavan 2006). Oxygen availability is an important factor during soil degradation. Aerobic degradation will result in the evolution of carbon dioxide and water. In the absence of oxygen, the organic matter will be converted into methane, carbon dioxide, and water (biogas). The schematic representation of soil degradation is shown in Fig. 8.3. This will result in weight loss and hence reducing wastes. These parameters can thus be used to determine the biodegradability rate (Leja and Lewandowicz 2010). Degradation percentage and time is taken for different types of nanocomposite by soil degradation method is given in the Tables 8.1 and 8.2.

Polyvinyl alcohol (PVA) and polycaprolactone (PCL) are the two synthetic polymers widely used to produce nanocomposites. PVA composites have a greater degradation percentage as compared to PCL composites. The degradation rate of PCL can be improved by incorporating bio nanoparticles than inorganic nanoparticles. Bio nanoparticles such as cellulose, starch, chitosan will increase the biodegradation rate as they are more prone to microbial attacks due to their more straightforward structure (Sarwar et al. 2018). Nanocellulose can be produced in different forms such as crystals, fiber, acylated CNF and are reported to increase biodegradability.

Chemical nanoparticles such as silver, titanium, copper, calcium carbonate have been reported for antibacterial and anti-fungal properties, thereby reducing the degradation rate. On the other side, natural polymers such as starch (thermoplastic starch), poly hydroxyl butyrate, and polylactic acid have a high degradation rate. So, these polymers can be added with chemical nanoparticles for preparing composite for various applications. Thermoplastic starch (TPS) added with different nanoparticles

Table 8.1 Degradation percentage and time for different nanocomposites

Nanocomposites	Degradation percentage (%)	Degradation time (days)	Refs.
PVA–CNP	98	20	Shruthy and Preetha (2019)
PVA–CNP–chitosan NP	70	150	Perumal et al. (2018)
PVA–starch (70%)	85	56	Azahari et al. (2011)
PVA–starch (50%)	40	56	Siddaramaiah and Somashekar (2004)
PVA–starch–TiO_2 NP	66	150	Hejri et al. (2013)
PVA–silver NP	86	50	Bari et al. (2016)
PVA–TPS–montmorillonite Na^+	75	72	Taghizadeh et al. (2011)
PLA–CNP	90	13	Luzi et al. (2016)
PLA–polybutylene succinate–CNP	90	17	Luzi et al. (2016)
PLA–TiO_2 NP	87	80	Kaseem et al. (2019)
PLA–cellulose–SiO NP	11	90	Lertphirun et al. (2019)
PLA–OMMT	83.3	80	Aguirre et al. (2018)
PLA–QAC	77.3	80	Aguirre et al. (2018)
PHB–silicate NC	100	50	Maiti et al. (2007)
PHB–silver NP	100	40	Castro Mayorga et al. (2017)
PHB–TiO_2 NP	67	32	Yew (2006)
PCL–TPS–CAB	30	112	Salah et al. (2014)
TPS–CNF	100	40	Mehran et al. (2015)
TPS–acylated CNF	100	60	Mehran et al. (2015)
TPS–montmorillonite NC	50	60	Behera (2018)
TPS–closite NC	45	60	Behera (2018)
TPS–egg shell NP	52	15	Bootklad and Kaewtatip (2013)
TPS–calcium carbonate NP	35	15	Bootklad and Kaewtatip (2013)

Abbreviations PVA—polyvinyl alcohol, PLA—poly lactic acid, PHB—polyhydroxy butyrate, PCL—polycaprolactone, TPS—thermoplastic starch, CNP—cellulose nanoparticle, NC—nanocomposite, NP—nanoparticles, CNF—cellulose nanofibers, CAB—cellulose acetate butyrate, OMMT—organo-modified montmorillonite, QAC—quaternary ammonium compounds

Table 8.2 Varation in degradation percentage with respect concentration of nano particles (Taghizadeh et al. 2011)

Composite	% of MMT-Na^+	Degradation %	Degradation days
PVA–TPS–MMT Na^+	0	64	72
PVA–TPS–MMT Na^+	1	67	72
PVA–TPS–MMT Na^+	3	70	72
PVA–TPS–MMT Na^+	5	75	72

Abbreviations PVA—polyvinyl alcohol, TPS—thermoplastic starch, MMT Na^+—montmorillonite

is reported to have the highest degradation (100%) in 20–60 days. Usage of eggshell nanoparticle has also been reported with a degradation rate of 52% in 15 days when incorporated in highly degradable TPS. Hence, it can also be used with other polymer composites. PVA being more hydrophilic as compared to others, is also reported to degrade at a faster rate (Cerqueira et al. 2017; Thakur et al. 2012; Kong et al. 2017).

8.3.1 Determination of Biodegradability by Measuring Carbon Dioxide Evolved

Composting is the most inexpensive and reliable test to evaluate the biodegradability of polymer composite in soil. The soil used can be checked for total dry solids, moisture content, volatile solids, pH, humidity, and microbial count. About 12 g carbon in the polymer/soil yields 44 g CO_2. CO_2 evolved from the compost without the bio nanocomposite can be kept as a control for such studies. It is also important to carry out such experiments in an environment devoid of CO_2 (Rosario and Dell 2010).

Test Under Controlled Compositing Conditions

According to 5338-98 ASTMD (American Society for Testing and Materials) standards, aerobic degradation of composite materials can be tested under controlled composting conditions. In order to carry out the test, a series of composting vessels can be taken, and it has to be filled with about 100 g testing soil and 10 g specimen (10:1 w/w). The sample specimen, along with positive and negative control, can be kept in triplicate (Iovino et al. 2008). The vessels should be incubated at 58 ± 2 °C, and the flow rate of air should be maintained at 10–40 mL/min. Quantification of CO_2 produced is done by placing 50 mL, 0.024 N $Ba(OH)_2$ in the chamber, which absorbs the CO_2 produced. Titration can be done against 0.05 N HCl to determine the absorbed CO_2. The analysis should be done at an interval of 3–4 days for the first 2–3 weeks, and then once in 1–3 weeks interval (Leejarkpai et al. 2011).

Where 'C test' is the amount of CO_2 evolved in the chamber with test specimens, C blank is the amount of CO_2 from a chamber with soil without a specimen,

$$\text{Biodegradation percentage} = (C\,test/C\,blank) * 100,$$

where C theoretical is the amount of CO_2 which can be produced by nanocomposite sample per molecule of carbon. It can be calculated using the equation;

C Theoretical = M* C* (44/12)

where 'M' is the mass of nanocomposite in gram, 'C' is the total organic carbon proportion in the nanocomposite taken, 12 and 44 are the atomic mass of carbon and molecular mass of carbon dioxide, respectively.

Test Under the Normal Composting Condition

According to 5988-03 ASTMD standards, aerobic biodegradation of composite plastic in the soil can be determined under normal room temperature conditions. The procedure is the same as that of ASTMD 5338-98 till the end of sample preparation. Incubation is done inside a chamber at room temperature. The CO_2 evolved here is measured by placing 50 mL, 0.05 N KOH solution inside the chamber. The absorbing solution can be titrated with 0.1 N HCl to measure the amount of CO_2 produced (Rudeekit et al. 2004).

Equipment Used for CO_2 Evolution Test

Different types of equipment (Fig. 8.4) used for the CO_2 evolution test are given below. It can be chosen based on the availability and the type of model setup prepared. It can be as simple as a glass jar method to complex MODA (Microbial Oxidative Degradation Analyzer) method.

Glass Jar Method

It is the low-cost method where a glass jar with a tight air lid can be employed for the test. The soil, along with the sample, can be placed in the jar. About 200 g of soil should be taken with a specimen weight of 10–20 g. Distilled water (50 mL) and 0.5 N KOH (20 mL) should be placed in the jar and close the lid (Rosario and Dell 2010).

Bio-meter Flask Method

Biometer flask assembly (250 mL) can be used to study the biodegradability. About 100 g of soil sieved through a 2 mm sieve and introduced to the flask along with

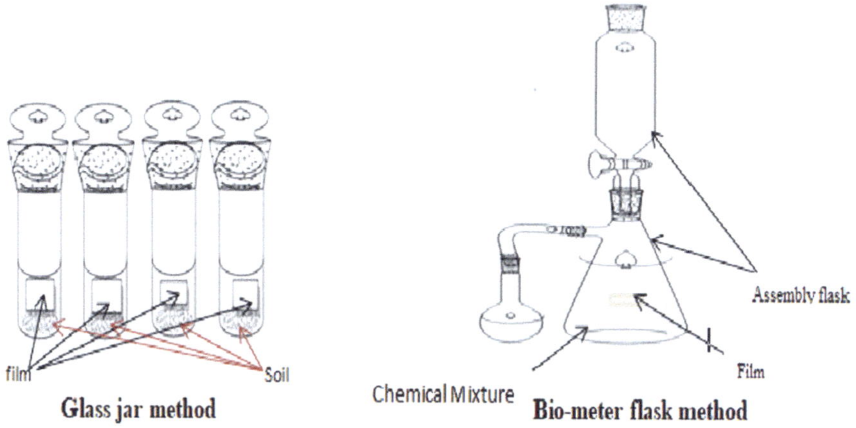

Fig. 8.4 Types of equipment used for CO_2 evolution test

a specimen of about 4–5 g. pH and moisture content of the soil should be between 6–8 and 50–70%, respectively. After adding the sample and soil, close the lids with a layer of glycerol. KOH can then be added via a syringe and can be kept for study. KOH can be removed at the end and titrated to know the amount of CO_2 produced and hence biodegradability (Rosario and Dell 2010).

MODA Method

Microbial Oxidative Degradation Analyzer (MODA) can be used to find out the degradation rate of bio nanocomposite based upon the amount of carbon dioxide produced. The analysis is based on the standard procedure as per ISO 14855-2 (Kunioka et al. 2007). The analyzer contains a reaction vessel which is maintained at uniform temperature and humidity throughout the reaction. These are controlled heater and by-passing humid air inside the reactor. The reactor can be filled with about 400 g sea sand mixed activated compost. 10 g of the composite sample can also be mixed with the sand, and then composting can be done. The reactor without the sample can be kept as control. The test can be done by maintaining the temperature at 58 °C at an airflow rate of 10 mL/min. Air given to the reactor is passed through a column of soda lime, which acts as a carbon dioxide trap followed by a column of water to maintain the humidity inside the reactor (Funabashi and Kunioka 2005; Funabashi et al. 2009). The gas produced upon degradation passes into the reactor, and the airflow, through a tube to a series of columns. These include ammonia trap (sulfuric acid), dehumidifying trap 1 (silicon dioxide), dehumidifying trap 2 (calcium chloride), absorption column for carbon dioxide, and absorption column for water. The carbon dioxide absorption column is filled with calcium hydroxide and calcium hydroxide solution, which can be titrated to find out the amount of CO_2 produced. Studies should be carried out for 30–130 days. Sand-compost-sample mixture can

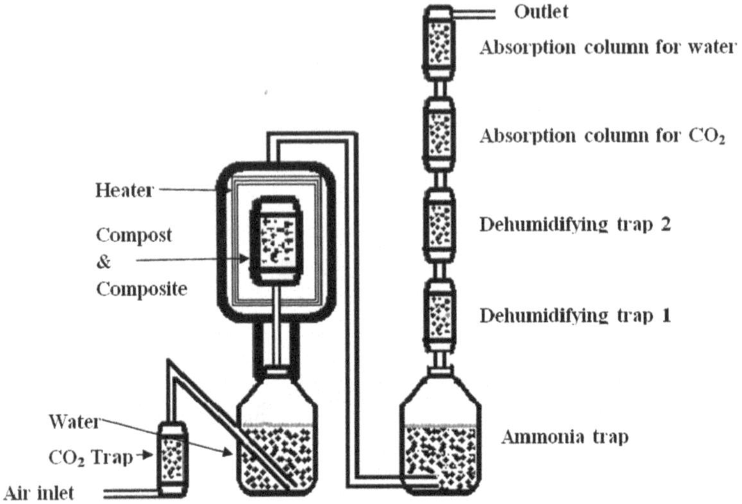

Fig. 8.5 Equipment for MODA

be mixed once a week during the study period, and the chemicals inside the CO_2 absorption column should be changed upon reaching 40% of absorption capacity. The MODA apparatus have depicted in Fig. 8.5 (Hoshino et al. 2007).

8.3.2 Determination of Biodegradability Under Anaerobic Condition

Under anaerobic conditions, polymers get converted into methane, carbon dioxide, and water. It is a usual practice to dump the waste deep inside the soil for degradation, where there is no oxygen availability. Hence, it is important to know the degradation rate of bio nanocomposite under anaerobic conditions. The test can be done according to ISO 14853 standards. The specimen, along with soil, is placed in a closed respirometer devoid of oxygen. During degradation, biogas will be generated inside and which increases the pressure inside the chamber. These pressure variations are measured using a manometer and are used to deduce the degradability rate. Ideal gas law can be used to find out the volume from the pressure value. Chromatography also can be done to analyze the composition of the gas generated (Massardier et al. 2005; Samešová et al. 2023). The biodegradation percentage can be calculated by the formula given below,

$$\text{Biodegradation\%} = (\text{Volume of biogas generated}/\text{Theoretical volume of biogas generated upon 100\%degradation}) * 100$$

8.3.3 Determination of Biodegradability by Measuring the Loss of Mass and Tensile Strength

Mass loss is a major physical change during composting. Mass loss occurs due to the conversion of organic matter in the polymer composite to carbon dioxide and water, which gets escaped and evaporated into the atmosphere. The mass loss will further reduce the mechanical and thermal properties of the polymer. Thus, polymer degradation can be analyzed by tracking the mass loss and loss of mechanical properties during the study period (Panthapulakkal and Sain 2012; Eckel et al. 2024). A plastic bin can be taken with tiny holes on the side and bottom to carry out the study to increase water and air circulation. About 500 g soil with 50% moisture and the weighed nanocomposite strip can be placed in the soil. Sample can be buried in the soil, along with positive and negative control (Fig. 8.6). Neat polymer strips without nanoparticles can be used as a positive control, and non-biodegradable polymer strips can be taken as a negative control (Obasi et al. 2013). Record the mass of the plastic bin after the preparation and monitor it weekly. If any weight loss is monitored during the study period, humidity should be maintained by spraying water at 50%. At the end of the study period, remove the soil sample and clean, rinse, and dry it. Record the mass of the strip when completely dried. The tensile strength and elongation properties of the strip can be tested after taking weight. Percentage reduction in tensile strength can be found from the test values recorded before and after the degradation study (Rosario and Dell 2010; Breslin 1993). The following equation can be used for calculating percentage biodegradation.

$$\text{Biodegradation\%} = (\mathit{Initialweight} - \mathit{Finalweight})/\mathit{Initialweight} * 100$$

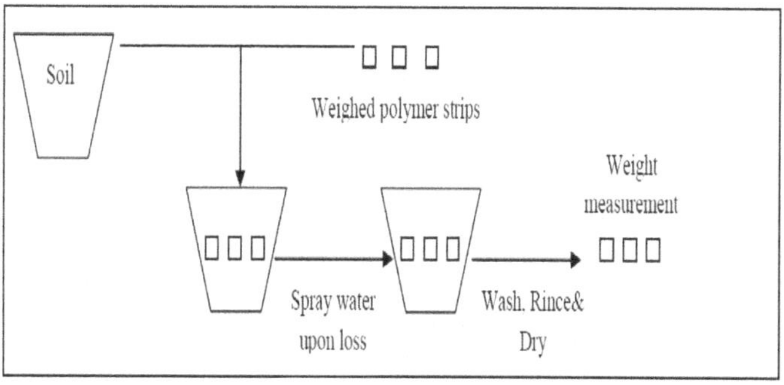

Fig. 8.6 Soil burial method (Shruthy and Preetha 2019)

8.3.4 Determination of Degradation by Size-Exclusion Chromatography

Size exclusion chromatography can be used to find out the degradation of nanocomposite upon composting or soil burial methods. The polydispersity index (PDI), weight average, and molecular weight can be found before and after degradation studies. Sample preparation for chromatography can be done by taking 20 mg of composite film dissolved in 10 cm^3 of tetrahydrofuran followed by filtration using 0.45 μm pore sized hydrophobic polytetrafluoroethylene. 100 μL of this sample can be injected for the chromatographic analysis. Polystyrene standards (0.5–2480 kDa) can be used to obtain the calibration curve (Aguirre et al. 2017).

8.3.5 Determination of Degradation by Disintegration Test

The test can be carried out according to ASTM D 5338 standard procedure. The test takes 12 weeks for completion. Test needed compost vessel filled with about 1200 g of compost and 200 g of the composite sample. Composite should be prepared in a dimension 2 cm $\times$ 2 cm squares and buried inside the compost. The compost needed to be passed through a 2 mm sieve before filling it into a vessel. The prepared vessels are then incubated at 580 $\pm$ 3 °C for 12 weeks. Water can be sprayed, or CO_2 free air can be passed through water to maintain the humidity of the compost. The vessel needed to be shaken and mixed once a week. This will provide uniform degradation over the specimen. Any change such as brittleness or cracks or holes occurring to the sample needed to be recorded weekly. At the end of the study period, the compost and the sample are sieved through a 2 mm screen. The amount of sample retained will be weighed and compared to the initial weight of the sample (200 g). If the sample retained is below 10%, the composite is disintegrable, and if the percentage retained is above 10%, the sample failed the test.

8.3.6 Plant Growth Study

It is done to conclude that the composite upon degradation does not cause any adverse effect on compost and plant growth. It is analyzed by percentage germination of the seed planted in the compost prepared and the bio nanocomposite polymer sample. Composite should be prepared in plastic pots without holes to prevent the loss of moisture content from the soil. Several vermiculture dilutions are also added due to the presence of excess salts and nutrients in the compost. OECD 208 guidelines can be followed for the test with modifications. Seeds will be counted when planted, and the germination percentage will be calculated. Biomass is determined by taking the average height of plants germinated. Biomass and germination percentage should be greater than or equal to 90% according to standards. Heavy metals in plants are also found to see whether the concentration is not crossing the acceptable levels.

8.4 Bacterial Degradation Studies

Naturally, derived materials will undergo complete degradation to complete the nutrient cycling in the ecosystem. The ubiquity of bacteria in the ecosystem will help with this cycling and can be considered a prerequisite for the degradation of bio nanocomposite. Few such bacteria reported are *Pseudomonas aeruginosa, Streptomyces species, Photobacterium fischeri, Agrobacterium tumefaciens, Vibrio harveyi,* and so on. Bacterial degradation can take place aerobically and anaerobically and under harsh conditions (Dong et al. 2002; Swift et al. 1979).

8.4.1 Determination of Degradation by BOD and COD Measurements

Different microbes have been isolated and been used for the faster degradation of the polymer-based nanocomposite. The degradability of the polymer can be evaluated by measuring the oxygen requirement for the degradation in a closed chamber with a respirometer. Specific microbes can be isolated and characterized from the soil and can be used to study the degradation rate (Kumar and Maiti 2016). The polymer composite can be cut into pieces and kept in an aqueous medium inoculated with the microbial sludge or selective microbes. The medium used is a mixture of phosphate buffer in water in the ratio of 5:1. About 250 mL medium, 100 mg polymer, and 10 mg inoculum can be taken for the studies. Upon incubation, microbes will multiply and will degrade the organic matter for energy needs. This will account for the respirometer to get the Biological Oxygen Demand (BOD) and Chemical Oxygen Demand (COD) measurements (Aguilar-Torrejón et al. 2023). The rate of biodegradation can be calculated from the BOD and COD values as follows.

$$\text{Biodegradation}\% = [BOD(mg/L/COD(mg/L)] * 100$$

8.4.2 Biofilm Assay by Spectroscopic Analysis

Biofilm formation is done by various microbes (bacteria, fungi, and protists) as an initial state of degradation. The microbes will attach to the composite upon exposure and form multiple layers over the composite. This will absorb more carbon and other sources from the nanocomposite surface where the film has been formed, leading to degradation. The composite ability to form a biofilm can be assessed by biofilm assay (Reynolds and Fink 2001). Composite used for the test should be sterilized by washing with 70% ethanol and irradiation by ultraviolet light. A microtiter plate with 24 wells can be used for the assay where the composite can be placed as four replicates. R2B broth and sterile compost extract (SCE) needed to be prepared for the assay (Merritt et al. 2011). Dry compost, rich in microbial flora, can be taken to prepare the compost extract. It is prepared by mixing 1:2 wt/vol, dry compost, and deionized water followed by two-minute vertexing. The supernatant of the mixture should be collected after allowing the mixture to settle for about 20 min. The supernatant is then filtered through a 1 mm mesh followed by passing it twice through a 0.22 μm mesh to prepare the sterile compost extract. Each well should be filled with 600 μL R2B and 200 μL SCE upon placing the sterile composite. Inoculated plates then needed to be incubated at 580 °C for 48 h with gentle shaking at 100 rpm. Biofilm forming bacteria such as *Pseudomonas aeruginosa* (PA) strain PAO1 can be used as a positive control, and wells without inoculation can be used as a negative control. The broth was removed from the plates and washed three times with water to analyze the biofilm formation. Plates are then added with about 800 μL 0.5% crystal violet for 15 min and washed three times with water followed by air drying. 800 μL 30% acetic acid is then added to the wells and incubated for 15 min. The microplate spectrometer is then used to analyze the absorbance at 600 nm of the incubated wells with the composite. Acetic acid is then decanted parallel to another microplate, and absorbance at 600 nm is taken. Quantification of biofilm formation is done by taking the difference between the average absorbance of cognate control and inoculated samples (Satti et al. 2017; Aguirre et al. 2018).

8.4.3 Biofilm Assay by SEM Analysis

Biofilm formation can also be analyzed by scanning electron microscopy. Composite to be analyzed should be placed in Erlenmeyer flask added with compost extract incubated overnight and R2B broth. Flask was then incubated at 58 °C for two days. Composite films can then be washed and air-dried to take the SEM micrographs (Aguirre et al. 2018).

8.4.4 Determination of Degradation by Percentage Weight Loss Measurement

Microbial degradation of bio nanocomposite can also be analyzed by weight loss due to organic matter utilization for microbial growth. A specific amount of composite can be kept inside a micro vial filled with a fixed amount of bacterial culture. The vial can then be incubated at a specific temperature for about two weeks. Vials should be prepared in sufficient number to record the weight loss intermittently during the study period (Ahmada et al. 2017; Mndlovu et al. 2024). The percentage of weight loss can be calculated as follows, which is directly proportional to the bacterial degradation rate.

$$\%\text{weight loss} = \mathit{Weight\ change\ after\ exposure\ to\ bacterial\ culture/Initial\ cumulative\ weight} * 100$$

8.4.5 *Bacterial Degradation Assay*

Biodegradability assay can also be done by the plating method. Nanocomposite should be cut into very thin films and should place on a coverslip. The media should be prepared based on the microbes used. Place the slips, add media on the plates, and inoculate the plate with specific bacterial culture suspended in the PBS (pH-7.4). Plates then needed to be incubated at 25 °C, and analysis can be done intermittently. A bright-field or fluorescence microscope can be used for the analysis. Food-borne pathogens such as *Listeria monocytogenes*, *Shigella flexneri*, *Salmonella typhi*, and *Staphylococcus aureus* can be used for inoculation and then identify the generation of toxic materials during the degradation. Growth inhibition of these pathogens signifies the toxic byproduct generation during degradation. After a specific time period, excessive growth of organisms signifies efficient biodegradation (Moghadam et al. 2011).

8.5 Other Degradation Studies

8.5.1 *Enzyme Degradation*

Microbes break down the polymers by producing enzymes, and these enzymes can be produced externally and added to polymers for faster degradation compared to microbial degradation. Testing can be done by placing the specimen in a medium with the required amount of enzymes. A CO_2 evolution test can be done to find out the degradation rate. Percentage biodegradation can also determine by analyzing the percentage weight loss of strength loss of polymer composite immersed in the enzyme (Spiridon et al. 2008; Shalem et al. 2024). Scanning electron microscopy (SEM) and atomic force microscopy (AFM) can be used to analyze the morphological changes during enzyme degradation. Fourier transform infrared spectroscopy (FTIR) can be done to analyze chemical composition changes during degradation. X-ray diffraction (XRD) and differential scanning calorimetry (DSC) can be done to check the crystallinity changes of the specimen during degradation. Enzyme first acts on the

Table 8.3 Variation in degradation percentage with respect concentration of nano particles (Taghizadeh et al. 2013)

Composite	% of MMT	Degradation %	Enzyme degradation Time (min)
PVA–S–CMC–MMT	0	34.54	4320
PVA–S–CMC–MMT	1	32.09	4320
PVA–S–CMC–MMT	3	21.57	4320
PVA–S–CMC–MMT	5	19.76	4320

Abbreviations PVA—polyvinyl alcohol, S—starch, CMC—carboxy methyl cellulose MMT—montmorillonite

surface of the composite film, then leading to the amorphous region and finally to the crystalline region (Table 8.3). The drawback of this method is the cost-effectiveness (Banerjee et al. 2014; Kumagai and Doi 1992).

8.5.2 Photodegradation

The photodegradation refers to the degradation of nanocomposite-based polymers under the application of ultraviolet or visible radiations. These radiations will break the polymer's chemical bonds and thus lead to the loss of various physical and mechanical properties. Photodegradation can occur aerobically by photo-oxidation, as well as anaerobically by chain breaking or cross-linking. Exposure to these radiations prior to soil or water degradation speeds up the rate of secondary degradation. In this method, the specimens are continuously exposed to radiation (UV/Visible) in a chamber for the study period, followed by various characterizations (Françaa et al. 2018). Polarized optical microscopy can be done to know morphological changes due to exposure. Optical microscopic images will be dense, thick, and non-porous for untreated films and will be having cracks upon radiation exposure. SEM analysis can also be done to check the surface morphology. DSC and XRD can be done to know the changes in the degree of crystallinity. FTIR can be done to check the chemical composition changes during exposure, and various mechanical changes can also be analyzed (Wellen et al. 2016).

8.5.3 Degradation by IR Radiation

Application of ionizing radiation to polymer nanocomposite causes chain scissoring leading to degradation. Structural modifications to the composite can be analyzed by measuring G-value using the solubility test. IR radiations can also cause cross-linking between polymer units leading to three-dimensional insoluble gel formation. The quantification of gel fraction from the solubility test can be found out from the ratio of the gel formed to the initial polymer weight, and the remaining fraction is the soluble fraction. Using Charles by-Pinner equation, a soluble fraction can be found out from the solubility test data.

$$\mathrm{S} + \mathrm{S}^{0.5} = (p/q) + (1/qPd)$$

where 'S' is the soluble fraction, 'p' and 'q' are the probabilities of chain scission and cross-linking per monomer unit per unit dose, 'P' is the average polymerization degree and 'd' is the absorbed dose (Kabanov et al. 2009).

Other techniques like swelling ratios, modulus of elasticity, light scattering, and nuclear magnetic resonance measurements can also be done to find out the degradation percentage. Molecular weight decline due to application of IR radiation can be studied by gel permeation chromatography and also by measuring intrinsic viscosity (Makuuchi and Cheng 2012).

8.5.4 *Mechanical Degradation*

Mechanical operations can be done to the bio nanocomposite to reduce its size. Reduction into small uniform size would lead to faster biodegradation. It can also be used as fillers. Regrinding is the most important method, which refers to the powdering process. Roll mill and ball mill are the most commonly used equipment for the process of regrinding. Ball milling can convert the size of the composite to below 100 μm. Size should be below 200 μm for the usage of the composite as a filler (Zevenhoven 2004).

8.5.5 *Oxidative Degradation*

It is the process of degradation by the reaction of oxygen with the polymer nanocomposite. Heat, UV, Photo radiations can stimulate oxidative degradation in the presence of oxygen. Degradation takes place due to the generation of free radicals and their reaction with the composite molecules. Oxidation thus leads to a loss in strength and utility. Mechanical properties can be analyzed before and after degradation. Oxidation also generates cracks and scratches over the surface of the composite leading to a reduction in weight and molecular weight. Other effects of oxidative degradation include a reduction in gloss and flexibility, breaking strain, impact resistance, and discoloration (Lannuzzi et al. 2013).

8.5.6 Chemical Degradation

It is the process of conversion of polymers to its monomers or oligomers by the action of various chemical agents. Based on agents used, types of degradation are categorized into five; hydrolysis, acidolysis, alcoholysis, aminolysis, and glycolysis. Changes and depolymerization of polymers when added to water are studied in alkaline, acidic, or neutral environments during hydrolytic degradation. Hydrolysis is the slowest process compared to other chemical degradation processes. The use of a catalyst such as an alkali metal hydroxide can speed up the degradation. Hydrolysis also needs a higher temperature and pressure. Acidolytic degradation can be done by treating the polymer composite with organic or inorganic acids. Some of the acids used include maleic acid, fumaric acid, and adipic acid. The process can be carried out at lower temperatures (60 °C) and in a short time period compared to hydrolysis. Alcoholytic degradation uses low molecular weight alcohol in the presence of catalysts. The rate of degradation is high for alcoholysis, even at low temperatures and shorter time periods. Glycolysis is a single-step conversion process, and which takes place with or without the action of a catalyst. Glycol added will transesterify the carbonate group of the polymer. It needs about 200–270 °C for 3–10 h. Potassium and sodium hydroxide can be used as a catalyst, reducing dissolution time (Achilias 2012).

8.5.7 Ultrasonic Wave Degradation

High-intensity ultrasound waves can cause a reduction in the molecular weight of the polymer nanocomposite. Composite can be ultrasonicated in suitable solvents for a specific time period to analyze the degradation rate by molecular weight determination. Sonication creates cavitation, which increases the pressure and temperature to a very high level. This causes the macromolecular cleavage by chain scission of the nanocomposite. Molecular weight reduction is directly proportional to the time of ultrasonication exposure and is also dependent on the solvent in which the composite is sonicated. It also leads to the generation of different types of free radicals, which will account for the degradation process (Price and Smith 1993; Basedow and Ebert 1977).

8.6 Conclusion

Applications of polymer-based biodegradable packaging are increasing day by day with innovations. Bio nanomaterials are proved to be incorporated in polymer to increase its strength and other properties, including thermal, chemical, and biological properties. Biodegradability is one of the most important reason for the increasing use of these polymers instead of normal plastics. Thus, it is important to study the biodegradability of the polymer nanocomposite. Soil biodegradability is one of the most important techniques as it is a common and economical way to dispose of waste. It is also important to analyze the degradation rate of products, toxicity of the degraded matter, and pollution impact in the atmosphere. This chapter discusses different types of biodegradation studies, and the information will be useful for conducting biodegradation studies for novel nanocomposite in water and soil under natural/artificial conditions.

References

Achilias DS (2012) Material recycling-trends and perspectives. InTech, Rijeka, Croatia. www.intechopen.com/books/material-recycling-trends-and-perspectives

Aguilar-Torrejón JA, Balderas-Hernández P, Roa-Morales G et al (2023) Relationship, importance, and development of analytical techniques: COD, BOD, and TOC in water—An overview through time. SN Appl Sci 5:118. https://doi.org/10.1007/s42452-023-05318-7

Aguirre EC, Auras R, Selke S, Rubino M, Marsh T (2018) Impact of nanoclays on the biodegradation of poly(lactic acid) nanocomposites. Polymers 10(202):1–22

Ahmada N, Sultanaa S, Azamb A, Sabira S, Khana MZ (2017) Novel bio-nanocomposites materials for enhanced biodegradability and photocatalytic activity. New J Chem 36:13, 1–28

Azahari NA, Othman N, Ismail H (2011) Biodegradation studies of polyvinyl-alcohol/corn starch blend films in solid and solution media. J Phys Sci 22:15–31

Banerjee A, Chatterjee K, Madras G (2014) Enzymatic degradation of polymers: a brief review. J. Mater Sci Technol 30(5):567–573

Bari SS, Chatterjee A, Mishra S (2016) Biodegradable polymer nanocomposites: An overview. J Polym Rev 56(2):287–328

Basedow AM, Ebert KH (1977) Ultrasonic degradation of polymers in solution. In: Physical chemistry. Springer, Berlin, Heidelberg, pp 83–148

Behera AK (2018) Mechanical and biodegradation analysis of thermoplastic starch reinforced nanobiocomposites. Mater Sci Eng 410(1):1–8

Bootklad M, Kaewtatip K (2013) Biodegradation of thermoplastic starch/eggshell powder composites. Carbohyd Polym 97:315–320

Breslin VT (1993) Degradation of starch-plastic composites in a municipal solid waste land fill. J Environ Polym Degrad 1–2:127–141

Briassoulis D, Dejean C (2010) Critical review of norms and standards for biodegradable agricultural plastics part I. Biodegrad Soil J Polym Environ 18:384–400

Briassoulis D, Pikasi A, Papardaki NG, Mistriotis A (2024) Biodegradation of plastics in the pelagic environment of the coastal zone—proposed test method under controlled laboratory conditions. Sci Total Environ 912:168889. https://doi.org/10.1016/j.scitotenv.2023.168889

Castro Aguirre, E Auras, R Selke, S Rubino, M. Marsh. Insights on the aerobic biodegradation of polymers by analysis of evolved carbon dioxide in simulated composting conditions. Polym. Degrad. Stab. 2017, 137, 251–271.

Castro Mayorga JL, Freitas F, Reis M, Prieto A, Lagaron JM (2017) Biosynthesis of silver nanoparticles and polyhydroxybutyrate nanocomposites of interest in antimicrobial applications. Int J Biol Macromol 108:426–435

Cerqueira JC, Da Silva Penha J, Oliveira RS, Guarieiro LLN, Da Silva Melo P, Viana JD, Machado BAS (2017) Production of biodegradable starch nano composite using cellulose nano crystals extracted from coconut fibres. J Polimeros 27(4):320–329

Cosgrove L, Geechan PL, Robson GD, Handley PS (2007) Fungal communities associated with degradation of polyester polyurethane in soil. J Appl Environ Microbiol 73:5817–5824

Dong J, Carey PR, Wei Y, Luo L, Lu X, Liu RQ, Dunaway-Mariano D (2002) Raman evidence for Meisenheimer complex formation in the hydrolysis reactions of 4-fluorobenzoyl- and 4-nitrobenzoyl-coenzyme a catalyzed by 4-chlorobenzoyl-coenzyme A dehalogenase. Biochemistry 41:7453–7463

Eckel F, Van Opdenbosch D, Schandl KS et al (2024) Dependency of tensile properties and biodegradation on molecular mass during hydrolysis of poly(butylene succinate). npj Mater Degrad 8:97. https://doi.org/10.1038/s41529-024-00512-3

EN 13432 (2000) Requirements for packaging recoverable through composting and biodegradation test scheme and evaluation criteria for the final acceptance of packaging

Françaa DC, Moraisa DD, Bezerraa EB, Araujoa EM, Wellen RMR (2018) Photodegradation mechanisms on poly(ε-caprolactone) (PCL). J Mater Res 21(5):1–8

Funabashi M, Kunioka M (2005) Biodegradable composites of poly(lactic acid) with cellulose fibers polymerized by aluminum triflate. Macromol Symp 224:309–321

Funabashi M, Ninomiya F, Kunioka M (2009) Biodegradability evaluation of polymers by ISO 14855-2. Int J Mol Sci 10:3635–3654

Hejri Z, Seifkordi AA, Ahmadpour A, Zebarjad SM, Maskooki A (2013) Biodegradable starch/ poly(vinyl alcohol) film reinforced with titanium dioxide nano particle. Int J Miner, Metall Mater 20-7:1–8

Hoshino A, Tsuji M, Momochi M, Sawada H, Kunioka M, Funabashi M, Uematsu S (2007) Study of the determination of the ultimate aerobic biodegradability of plastic materials under controlled composting conditions method by gravimetric method of evolved carbon dioxide using microbial oxidative degradation analyzer (MODA). J Polym Environ 15:275–280

International Standardisation, ISO 14853 (1999) Determination of the ultimateanaerobic biodegradability of plastic materials in an aqueous system—method by measurement of biogas production

International Standardisation, ISO 14851 (1999) Determination of the ultimate aerobic biodegradability of plastic materials in an aqueous medium—method by measuring the oxygen demand in a closed respirometer

International Standardisation, ISO 14855-1 (2004) Determination of ultimate aerobic biodegradability of plasticmaterials under controlled composting conditions—method by analysis of evolved carbon dioxide

Iovino R, Zullo R, Rao MA, Cassar L, Gianfreda L (2008) Biodegradation of poly(lactic acid)/starch/ coir biocomposites under controlled composting conditions. Polym Degrad Stab 93:147–157

Kabanov VYa, Feldman VI, Ershov BG, Polikarpov AI, Kiryukhin DP, Apel PYu (2009) Radiation chemistry of polymers. J High Energy Chem 43(1):1–18

Kale G, Kijchavengkul T, Auras R, Rubino M, Selke SE, Singh SP (2007) Compostability of bioplastic packaging materials: an overview. Macromol Biosci 7(3):255–277

Kaseem M, Hamad K, Rehman Z (2019) Review of recent advances in polylactic acid/TiO_2 composites. J Mater 12–3659:1–16

Kjeldsen A, Price M, Lilley C, Guzniczak E. A review of standards for biodegradable plastics. Industrial Biotechnology Innovation Centre, pp 1–28

Kong J, Yu Y, Pei XP, Han C, Tan Y, Dong L (2017) Polycaprolactone nanocomposite reinforced by bio resource starch-based nanoparticles. Int J Biol Macromol 102:1304–1311

Kumagai Y, Doi Y (1992) Enzymatic degradation and morphologies of binary blends of microbial poly(3-hydroxy butyrate) with poly(e-caprolactone), poly(1,4-butylene adipate and poly(vinyl acetate). Polym Degrad Stab 36:241–248

Kumar S, Maiti P (2016) Controlled biodegradation of polymers using nanoparticles and their applications. J RSC Adv 72:1–96

Kunioka M, Ninomiya F, Funabashi M (2007) Novel evaluation method of biodegradabilities for oil based poly caprolactone by naturally occurring radiocarbon-14 concentration using accelerator mass spectrometry based on ISO 14855–2 in controlled compost. Polym Degrad Stab 92:1279–1288

Lannuzzi G, Mattsson B, Rigdahl M (2013) Color changes due to thermal ageing and artificial weathering of pigmented and textured ABS. J Polym Eng Sci 53(8):1687–1695
Leejarkpai T, Suwanmanee U, Rudeekit Y, Mungcharoen T (2011) Biodegradable kinetics of plastics under controlled composting conditions. J Waste Manag 31:1153–1161
Leja K, Lewandowicz G (2010) Polymer biodegradation and biodegradable polymers—a review. Polish J Environ Stud 19:255–266
Lertphirun K, Kaweesrikulki (2019) Properties of poly lactic acid filled with hydrophobic cellulose/ SiO_2 composites. Int J Polym Sci 1–8
Liang J, Shu M-Y, Huang H-R, Ma C-W, Kim S-K (2024) Using benthic indices to assess the ecological quality of sandy beaches and the impact of urbanisation on sandy beach ecosystems. J Mar Sci Eng 12:487. https://doi.org/10.3390/jmse12030487
Luzi F, Fortunati E, Jiménez A, Puglia D, Pezzolla D, Gigliotti G, Kenny JM, Chiralt A, Torrea L (2016) Production and characterization of PLA PBS biodegradable blends reinforced with cellulose nano crystals extracted from hemp fibres. Ind Crop Prod 93:276–289
Maiti P, Batt CA, Giannelis EP (2007) New biodegradable poly hydroxybutyrate/layered silicate nano composites. J Bio Macromol 8–11:3393–3400
Makuuchi K, Cheng S (2012) Radiation processing of polymeric materials and its applications. Wiley, pp 1–26
Massardier Nageotte V, Pestre C, Pradet TC, Bayard R (2006) Aerobic and anaerobic biodegradability of polymer films andphysico-chemical characterization. Polym Degrad Stab 91:620–627
Merritt JH, Kadouri DE, Toole GAO (2011) Growing and analyzing static biofilms. Curr Protoc Microbiol 1–18
Mndlovu H, Kumar P, du Toit LC et al (2024) A review of *biomaterial* degradation assessment approaches employed in the biomedical field. *npj Mater Degrad* 8:66. https://doi.org/10.1038/s41529-024-00487-1
Moghadam PN, Azariyan E, Sharifia I (2011) Conductive and biodegradable polyaniline/starch blends and their composites with polystyrene. Iran Polym J 20:319–328
Narayn R, Pettigrew CA (1999) ASTM standards help define and grow a new degradable plastics industry. ASTM Stand News 27(12)
Obasi HC, Igwe IO, Madufor IC (2013) Effect of soil burial on tensile properties of polypropylene/ plasticized cassava starch blends. J Adv Mater Sci Eng, J At mol 3(3):530–536
Panthapulakkal S, Sain M (2012) Preparation and characterization of cellulose nanofibril films from wood fibre and their thermoplastic polycarbonate composites. Int J Polym Sci 1–6
Perumal AB, Sellamuthu PS, Nambiar RB, Sadiku ER (2018) Development of polyvinyl alcohol/ chitosan bio-nano composite films reinforced with cellulose nano crystals isolated from rice straw. Appl Surf Sci 449:591–602
Price GJ, Smith PF (1993) Ultrasonic degradation of polymer solutions—III. The effect of changing solvent and solution concentration. Eur Polym J 29(2, 3):419–424
Raghavan D (2006) Charecterization of biodegradable plastics. J Polym-Plast Technol Eng 34–1:41–63
Reynolds TB, Fink GR (2001) Baker's yeast, a model for fungal biofilm formation. Science 291:878–881
Rosario L, Dell E (2010) Biodegradability of plastics testing in an undergraduate materials laboratory course. Am Soc Eng Educ 15(231):1–12
Rudeekit Y, Leejakpai T, Euaphantasate N, Kongsuwan K (2004) Comparative degradation of biodegradable plastics by aerobic microorganism using ASTMD 5338-98(03) and ASTMD 5988-03 as standard methods, vol 10
Rutkowska M, Jastrzebska M, Janik H (1998) Biodegradation of poly caprolactone in sea water. React Funct Polym 38:27–30
Rutkowska M, Heimowska A, Krasowska K, Janik H (2002) Biodegradability of poly ethylene starch blends in sea water. Pol J Environ Stud 11:267–274
Salah F, Ali A (2014) Biodegradation properties of poly-e-caprolactone, starch and cellulose acetate butyrate composites. J Polym Environ 22:359–364

Samešová D, Poništ J, Hybská H et al (2023) Determination of aerobic and anaerobic biological degradability of waste tyres. Environ Sci Eur 35:13. https://doi.org/10.1186/s12302-023-00723-6

Sarwar MS, Niazi MBK, Jahan Z, Ahmad T, Hussain A (2018) Preparation and characterization of PVA/nanocellulose/Ag nano composite films for antimicrobial food packaging. J Carbohydarte Polym 184:453–464

Satti SM, Shah AA, Auras R, Marsh TL (2017) Isolation and characterization of bacteria capable of degrading poly(lactic acid) at ambient temperature. Polym Degrad Stab 144:392–400

Shalem A, Yehezkeli O, Fishman A (2024) Enzymatic degradation of polylactic acid (PLA). Appl Microbiol Biotechnol 108:413. https://doi.org/10.1007/s00253-024-13212-4

Shruthy R, Preetha R (2019) Cellulose nanoparticles from agro-industrial waste for the development of active packaging. Appl Surf Sci 484:1274–1281

Siddaramaiah, Raj B, Somashekar R (2004) Structure-property relation in polyvinyl alcohol/starch composites. J Appl Polym Sci 91:630–635

Spadaro G, Alessi S, Dispenza C Ionizing radiation-induced cross linking and degradation of polymers. In: Applications of ionizing radiation in materials processing. Università degli Studi di Palermo, Italy, Ch 7, pp 167–182

Spiridon I, Popescu MC, Bodârlău R, Vasile C (2008) Enzymatic degradation of some nanocomposites of poly(vinyl alcohol) with starch. Polym Degrad Stab 93:1884–1890

Swift MJ, Heal OW, Anderson JM (1979) Decomposition in terrestrial ecosystems. Stud Ecol 5

Taghizadeh MT, Sabouri N, Ghanbarzadeh B (2013) Polyvinyl alcohol:starch:carboxy methyl cellulose containing sodium montmorillonite clay blends; mechanical properties and biodegradation behavior. J SpringerPlus 2(376):1–8

Taghizadeh MT, Abbasi Z, Nasrollahzade Z (2011) Study of enzymatic degradation and water absorption of nanocomposites starch/polyvinyl alcohol and sodium montmorillonite clay. J Taiwan Inst Chem Eng 43:120–124

Takishita K, Tsuchiya M, Reimer JD, Maruyama T (2006) Molecular evidence demonstrating the basidiomycetous fungus Cryptococcus curvatus is the dominant microbial eukaryote in sediment at the Kuroshima Knoll methane seep. Extremophiles 10:165–169

Thakur VK, Singha AS, Thakur MK (2012) Green composites from natural fibers, mechanical and chemical aging properties. Int J Polym Anal Charact 17(6):401–407

Tosin M, Weber M, Siotto M, Lott C, Degli IF (2012) Laboratory test methods to determine the degradation of plastics in marine environmental conditions. Front Microbiol 3:225

Tsuji H, Suzuyoshi K (2002) Environmental degradation of biodegradable polyesters poly(caprolactone), poly[(R)-3-hydroxybutyrate] and poly(L-lactide) films in natural dynamic sea water. J Polym Degrad Stab 75:357–365

Weber M, Lott C, Fabricius KE (2006) Sedimentation stress in ascleractinian coral exposed to terrestrial and marine sediments with contrasting physical, organic and geochemical properties. J Exp Mar Biol Ecol 336:18–32

Wellen RMR, Canedo EL, Fechine GJM (2016) Photodegradation and photostabilization of poly(3-hydroxybutyrate). J Mater Res 19–4:759–764

Yew S-P, Tang HY, Sudesh K (2006) Photocatalytic activity and biodegradation of polyhydroxybutyrate films containing titanium dioxide. Polym Degrad Stab 91:1800–1807

Zevenhoven R (2004) Treatment and disposal of polyurethane wastes: options for recovery and recycling. Helsinki University of Technology, Espoo, pp 1–50

Chapter 9
Exploring the Role of Magnetic Nanoparticles in Cancer Imaging and Diagnostic Advancements

Zehra Durmus and Sumeyra Coskun

Abstract This chapter summarizes the recent developments in structures, properties, and applications of superparamagnetic iron oxide nanoparticles (SPIONs). It aims to explore how superparamagnetic iron oxide nanoparticles (SPIONs) can significantly improve the quality of magnetic resonance imaging (MRI) by acting as effective contrast agents, thereby aiding in the early diagnosis and monitoring of tumors. It discusses the application of SPIONs in magnetofection, where their magnetic properties are utilized to enhance the efficiency of gene delivery into target cells, potentially revolutionizing gene therapy and diagnosis. The chapter highlights the use of SPIONs in magnetic separation techniques, emphasizing their ability to quickly and efficiently isolate specific cell populations from complex mixtures, which is crucial for biomedical research and clinical diagnostics. It addresses the role of SPIONs in cell labeling for in vivo studies, enabling researchers to monitor cell behavior and migration using MRI, thus providing insights into cellular processes relevant to disease progression and treatment efficacy via imaging and diagnosis.

Keywords Magnetic nanoparticles · Cancer · SPION · Diagnosis · Imaging

9.1 Applications of Magnetic Nanoparticles for Medical Nanobiotechnology

People have been exploring magical ways to reduce the pain of illness and injury for hundreds of years. Numerous scholars believe that accomplishing this objective may depend on the application of nanotechnology in medicine. These applications use molecularly-level nanostructures and nanodevices to generate, monitor,

Z. Durmus (✉)
Department of Biomedical Engineering, Faculty of Engineering and Natural Sciences, Biruni University, 34015 Istanbul, Türkiye
e-mail: zehrad@biruni.edu.tr

S. Coskun
Faculty of Pharmacy, Bezmialem Vakif University, 34093 Istanbul, Türkiye

P. M. Visakh (ed.), *Engineering Applications of Polymer-Based Nanocomposites*, Advanced Structured Materials 237, https://doi.org/10.1007/978-3-031-93463-6_9

control, repair, and safeguard entire biological human systems. Nanotechnology has the potential to revolutionize medical research and open up new fields in human development, there is a strong road for using nanotechnology in medicinal, preventative, and diagnostic applications. Nanotechnology plays a pivotal role in contemporary medicine by manipulating materials at the nanoscale, which is essential for cell scale to comprehend signaling, enzyme action, and cycles (Roeder 2013). This technology offers tools to investigate tissue structure at the nanoscale, aiding in understanding biological processes between atomic and cellular levels (Mura et al. 2013). Moreover, it facilitates the fabrication of synthetic biomaterials for therapeutic purposes and tissue replacement (Dobrovolskaia and Mcneil 2007). One of the most promising applications of nanotechnology in medicine is targeted drug delivery. Nano drugs can be customized to fit specific disease-associated sites precisely, ensuring accurate treatment while minimizing harm to healthy cells (Wilhelm et al. 2016). Nanotechnology also enables the transportation of delicate organic macromolecules and peptides to their intended sites, shielding them from degradation and immune responses. It assists in surmounting barriers that typically obstruct the passage of large molecules. Furthermore, nanotechnology has transformed medical diagnostics, providing more cost-effective and convenient disease detection methods. Beyond diagnostics and therapeutics, nanotechnology aids in managing diseases that demand strict drug regimens by automating the release of medications, thereby enhancing patient compliance. Overall, nanotechnology has the potential to significantly enhance medical diagnostics, treatment, and prevention, offering innovative tools for addressing diseases at the nanoscale.

The Nobel Prize-winning physicist Richard P. Feynman created the concept of nanotechnology, saying that "there is plenty of room at the bottom" to sign for the production of more physical things and mechanisms at the scale of tiny tools and particles. Nowadays, not only physicists but also scientists from diverse disciplines believe that forthcoming nanofabrication technologies and instruments, such nanomachines, robotics, nanomedicine, and diagnostic gadgets, would fetch in remarkable biological marvels for the medical field and other sectors (Thakur et al. 2021). Given that nanotechnology is viewed by scientists as the science of the future, they anticipate reaping several benefits from its implementation. Additionally, medical professionals, researchers studying biology, and clinicians have been interested in the distinct qualities and behaviors of nanoscale items.

Since then, with the lure of promised opportunities nanotechnology has extensive applications as nanomedicine in the medical field. Some nanoparticles have possible applications in diagnostics, imaging, and therapy/treatment such as; targeted medicinal products, pharmaceutical products, biomedical implants, and tissue engineering. Biomedicine also now stands at another threshold with the full potential of nanobiotechnology, with the use of artificial intelligence for predictive designs, assays and data results and treatments. In this chapter, different types of nanoparticles will be discussed used in the medicine and they will be explained applications of that nanoparticles in diagnostics and imaging (Coskun 2016).

Cancer

Cancer is defined as a group of disorders where uncontrolled cells proliferate uncontrollably and have the ability to spread to neighboring tissues. An uncontrollably growing collection of cells deviates from the standard rules of cell division. Physiological signals for cell division, differentiation, or death are continuously applied to normal cells. The cancer cells can spread and proliferate continuously, which might be lethal (see Fig. 9.1). In conclusion, metastasis—the process by which tumors spread—is responsible for about 90% of cancer-related deaths. Cancer is a complicated collection of diseases with numerous potential causes, such as genetics, lifestyle choices like food, exercise, and tobacco use, infections, and exposure to various chemicals and radiation in the environment. The molecular networks governing cell division, proliferation, and death are similar in all mammalian cells. These networks undergo alterations at the molecular, biochemical, and cellular levels when healthy cells develop into cancerous cells. Cancer generally involves alterations or mutations in the genome of the cell. Proteins created by DNA mutations upset the delicate balance between cell division and quiescence.

Cancer cells can also spread to other parts of the body through the blood and lymph systems. There are several main types of cancer. Carcinoma is a cancer that begins in the skin or in tissues that line or cover internal organs. Sarcoma is a cancer that begins in bone, cartilage, fat, muscle, blood vessels, or other connective or supportive tissue. Leukemia is a cancer that begins in blood-forming tissue, such as the bone marrow, and causes too many abnormal blood cells to be made. Lymphoma and multiple myeloma are cancers that begin in the cells of the immune system. Central nervous system cancers are cancers that begin in the tissues of the brain and spinal cord which also called malignancy. Some patients with metastatic cancer may live longer with treatment. However, controlling the cancer's progress or alleviating its symptoms is usually the main objective of therapy for metastatic cancer. The majority of cancer patients die from metastatic disease, and metastatic tumors can seriously impair bodily functions. The World Health Organization (WHO) reports that the

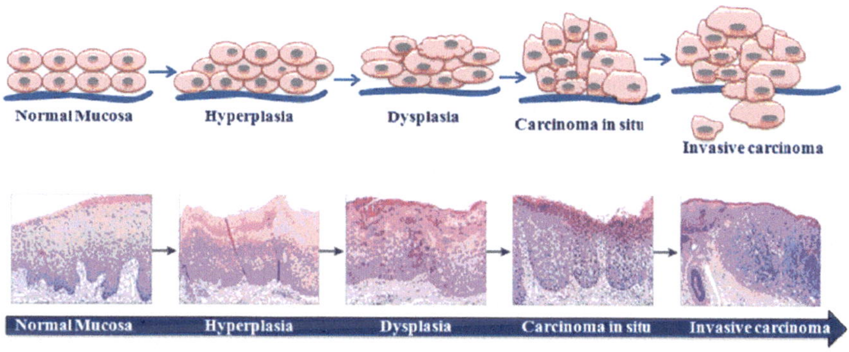

Fig. 9.1 Stages of cancer progression. Reproduced with permission from Shah et al. (2011). Copyright (2011) Springer

cancer burden is still rising worldwide, placing a great deal of financial, emotional, and physical strain on people, families, communities, and healthcare systems. Many cancer patients worldwide lack access to prompt, high-quality diagnosis and treatment, and many health systems in low- and middle-income nations are ill-equipped to handle this load. Accessible early detection, high-quality treatment, and survivorship care are helping many cancer types survive in nations with robust health systems.

In recent years, numerous studies have been conducted to investigate the clinical applications of nanotechnology for the detection, diagnosis and treatment of cancer and other diseases, and efforts have been increasing to develop a variety of organic or inorganic nanoparticles, which can be categorized as different nanoscale materials such as quantum dots, gold nanoparticles, iron oxide nanoparticles, up-conversion nanophosphors, metal organic lattices and zeolites, and purely inorganic or purely organic materials such as liposomes (Gao et al. 2004; Huang et al. 2006).

Many of the nanomaterials in these studies show promise in biomedical and preclinical applications due to their tunable size, high active substance transport capacity, functionalizable surface chemistry, controllable drug release kinetics, improved biocompatibility and passive or active tumor targeting, and enhanced permeability and retention (EPR) properties (Davis et al. 2008; Kim et al. 2013). As can be seen in the following cited Table 9.1 from Tian et al. study and Clemente's thesis work, nanoparticles with different chemical and structural properties enable the application of highly effective imaging techniques such as magnetic resonance imaging (MRI), positron emission tomography (PET), computed tomography (CT), single photon emission computed tomography (SPECT), and optical imaging (Clemente 2018; Tian and Zhong 2013). Nanoparticles are also used in various cancer-fighting therapies and applications such as chemotherapy, photodynamic therapy, neutron capture therapy, thermal therapy, and magnetic therapy (Bardhan et al. 2011; He et al. 2015; Petros and DeSimone 2010).

9.2 Magnetic Nanoparticles for Cancer Imaging and Diagnostics

In the 1990s, clinical molecular imaging emerged as a promising and minimally invasive in vivo tool to visualize cellular functions and monitor cells with the perception that the underlying causes of most modern diseases could only be understood through increasingly specific cellular or subcellular studies. In recent years, advancements in biomedical imaging technologies have significantly expanded the capabilities of existing modalities. Concurrently, extensive scientific research has leveraged these diverse imaging techniques to investigate novel aspects of the tumor microenvironment and enhance the assessment of established biomarkers. Consequently, there is a crucial need to examine the current state-of-the-art methods for characterizing tumors and their microenvironment in vivo through imaging.

Table 9.1 Characteristics of the most popular molecular imaging techniques, both pre-clinical and clinical (Clemente 2018; Tian and Zhong 2013)

Technique	Energy used	Common contrast agents	Imaging agent mass	Spatial resolution (mm)	Temporal resolution	Sensitivity of detection (mol L^{-1})	Penetration depth (mm)	Quantification efficiency
CT	X-rays	I or Ba agents	mg range	0.03–1	Minutes	0.1	>300	–
SPECT	γ-rays	γ-emitters	<1 ng	0.5–15	Minutes	10^{-11}	>300	High
PET	Annihilation photons	β^+ emitters	<1 ng	1–10	Seconds–minutes	10^{-12}	>300	High
MRI	Radio frequency waves	Gd^{3+} agents or Fe_3O_4 nanoparticles	0.01–100 mg	0.03–1	Minutes–hours	10^{-5}	>300	Medium
Optical	Visible to infrared light	Fluorophores or lanthanides	<1 ng–10 mg	2–10	Seconds–minutes	10^{-11}	1–20	Medium
Ultrasound	High frequency sound waves	Gas microbubbles	0.01–10 mg	0.05–1	Seconds–minutes	10^{-8}	1–200	Low
CT	X-rays	I or Ba agents	mg range	0.03–1	Minutes	0.1	>300	––

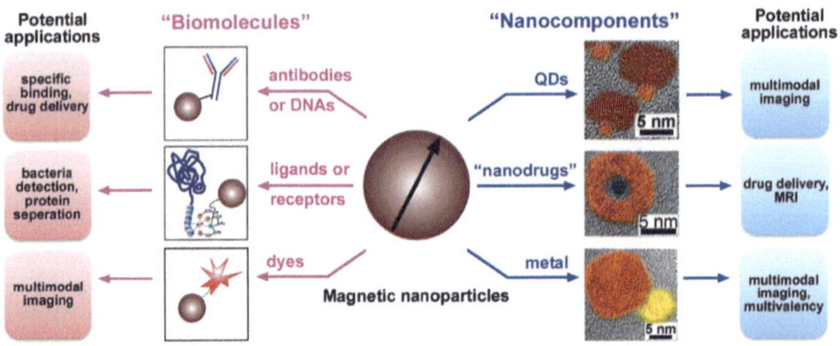

Fig. 9.2 Iron oxides (Fe_2O_3 and Fe_3O_4), metal ferrites (MFe_2O_4, M = Cu, Co and Ni), Fe(0), Co(0), Ni(0), and multi-component nanoparticles. Reproduced with permission from Bárcena et al. (2009), Gao et al. (2009). Copyright (2009) Springer and American Chemical Society

As nanobiotechnology has advanced recently, a wide variety of carbon-based nanoparticles, including carbon nanotubes (CNTs), fullerene, graphene, lipid, liposome, polymer, dendrimer, micelle, biological, and magnetic nanoparticles (shown in Fig. 9.2), have been employed as imaging and diagnostic materials. The use of magnetic nanoparticles (MNPs) in biomedical applications, including magnetic resonance imaging (MRI), virus detection, tissue repair, immunoassay, biological fluid detoxification, magnetic cell separation, enzyme catalysis, gene therapy, hyperthermia, targeting chemotherapy, and radiotherapy, has drawn more and more attention.

Small size, vast surface area, in vivo drug administration, and unique electronic, optical, and magnetic capabilities are some of the pharmacokinetic and physical characteristics of nanoparticles, but they are also beneficial. Thus, as nanotechnology has advanced, many multifunctional nanometarials have also appeared, with a particular emphasis on healthcare applications. Creating novel methods to lessen side effects associated with both conventional diagnosis and treatment is one of the largest issues facing medical science (Hosu et al. 2019).

Nanoparticles are thought to be beneficial for cancer detection and treatment because they can serve as drug carriers, imaging contrast agents, and can be adorned with biorecognition ligands for cancer tissue targeting, as shown in Fig. 9.2 (Gao et al. 2009). The nanoparticles are small enough to get past the mucosal barrier and reach the cancer target, which is the cytoplasm or nucleus of the cell. Recent studies have shown that various functional nanoparticles can be used as drugs or targeting moieties for effective and site-specific delivery when covalently bound to a range of biological molecules, such as peptides, proteins, nucleic acids, or small molecule ligands (Henglein 1989). Biorecognition ligands have the ability to guide nanoparticles toward cancer targets. External stimuli that have a noncontact affinity for nanoparticles can also be used to induce it.

Superparamagnetic iron oxide nanoparticles, or SPIONs, are great imaging agents that are employed in the diagnosis and treatment of cancer (in Fig. 9.3). Many applications of SPIONs in biotechnology have emerged in recent years using SPIONs

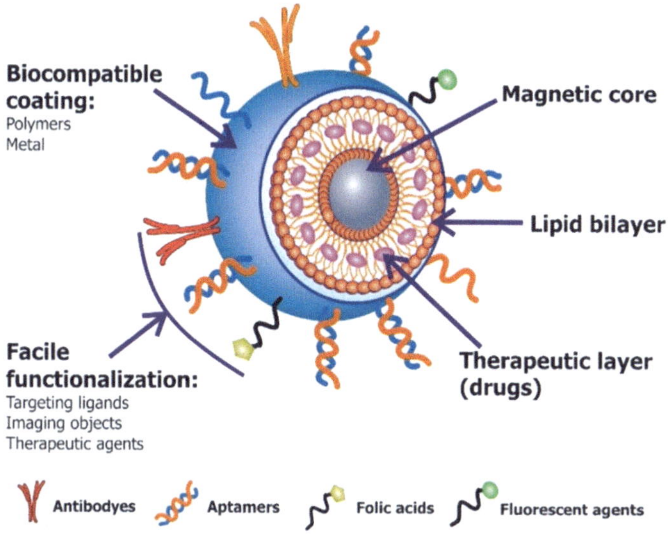

Fig. 9.3 Schematic illustration of a multifunctional magnetic nanoparticle structure with different types of coatings, target ligands and imaging agents. Therapeutic drugs can be embedded in the coating or conjugated on the surface. Reproduced with permission from Belyanina et al. (2017). Copyright (2017) MDPI

as an MRI contrast agent for cancer diagnostics is among its most intriguing uses (Ferrari 2005). They have several benefits, are metabolized by the body, and lower the risk of chronic cytotoxicity (Mahmoudi et al. 2011). The capacity of SPIONs to accumulate drugs at specific sites and create hyperthermia for cancer treatment by external magnetic manipulation makes them useful not only for imaging and diagnostics but also for therapy. To identify malignancy, SPIONs are employed as a contrast agent (Zhao et al. 2013). According to appearance, the most recent advancements in nanotechnology connected to cancer therapy involve a combined strategy in which a cancer will be identified, diagnosed, and targeted for treatment utilizing the same delivery system.

9.2.1 Magnetic Resonance Imaging (MRI)

MRI is a noninvasive medical imaging technique, CT, SPECT, and PET, It avoids the use of ionizing electromagnetic radiation and the associated hazards (the energy emitted can interact and disrupt chemical bonds within living tissue (Reisz et al. 2014), that physicians use to diagnosis and whole medical treatment (Tran and Webster 2010). MRI is related to the powerful magnetic field, radiofrequency (RF), a computer to produce detailed pictures of organs, soft tissues, bone, and virtually all other internal body structures. All hydrogen atoms have a nuclear spin, which is

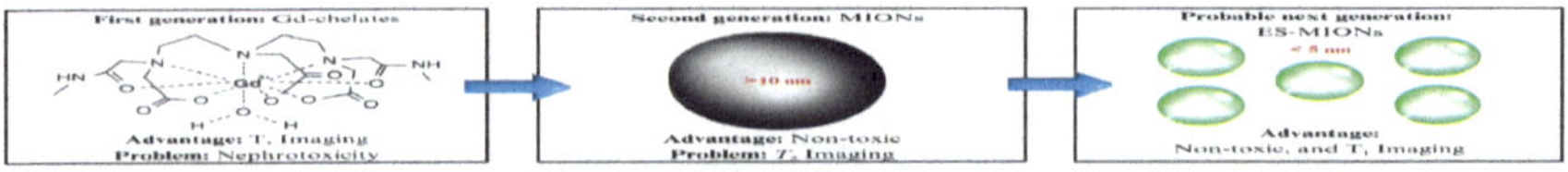

Fig. 9.4 A summary of the advantages and disadvantages of first, second and the probable subsequent generations of MRI contrast agents. MIONs: magnetic iron oxide nanoparticles; ES-MIONs: extremely small MIONs. Adapted with permission from Shen et al. (2017). Copyright (2017) American Chemical Society

related to a magnetic dipole moment, and are thus detectable by MRI. MRI scanners consist of two powerful magnets. Also, the human body is seventy percent water molecules, which consist of hydrogen and oxygen atoms. At each atom center have a proton, which operates as a magnet and is sensitive to any magnetic field (Fig. 9.4).

In our body, normally the water molecules are randomly organized, but when entering an MRI scanner, the first magnet causes the body's water molecules to align in one direction, either north or south, and the second magnet is turned on and off in quick pulses series, causing each hydrogen atom to alter its alignment and then when switched off, quickly switch back to its original relaxed state. The scanner can detect them and, contacts with a computer, lastly, creates a detailed cross-sectional image. Other parameters are the Néel Relaxation and the Np anisotropy also play a key role in MRI working principle (Fig. 9.5).

To force the spins of hydrogen nuclei from the body to align both parallel and antiparallel (as determined by the Boltzmann distribution), MRI generates an external strong magnetic field, typically between 0.5 and 3.0 T. This results in a new overall magnetic moment oriented along the axis of the MRI scanner. Using radiofrequency pulses with preset repetition times, extra energy is injected momentarily, causing the proton to spin in the opposite direction of the externally applied magnetic field during the duration of the pulse. Relaxivity is the mechanism by which the proton spins realign with the magnetic field axis after the radiofrequency pulse stops (Fig. 9.6) (Clemente 2018).

Two distinct processes are typically used to assess this relaxivity: T1 measures the longitudinal relaxivity's recovery in the z axis direction, and T2 measures the transverse magnetization's gradual disappearance inside the x, y plane. Since T1 typically lasts 1 s and T2 lasts 100 ms, T2 weighted investigations are more sensitive and appropriate for molecular or functional metabolic imaging (Currie et al. 2013; Shen et al. 2017).

Magnetic resonance imaging is characterized by *the spin–lattice relaxation time*, that time contrast known as **T1**. *The spin–spin relaxation time* is known as **T2**. The spin–spin relaxation time is referred to as transverse relaxation or **T2** and describes the decay of the excited magnetization perpendicular to the applied magnetic field. The spin–lattice relaxation time is referred to as longitudinal relaxation or **T1** and describes the return to equilibrium in the direction of the magnetic field (Jeon et al. 2021; Plewes and Kucharczyk 2012). A variety of signals can be produced, detected, analyzed, and reconstructed as greyscale cross-sectional images by using gradient

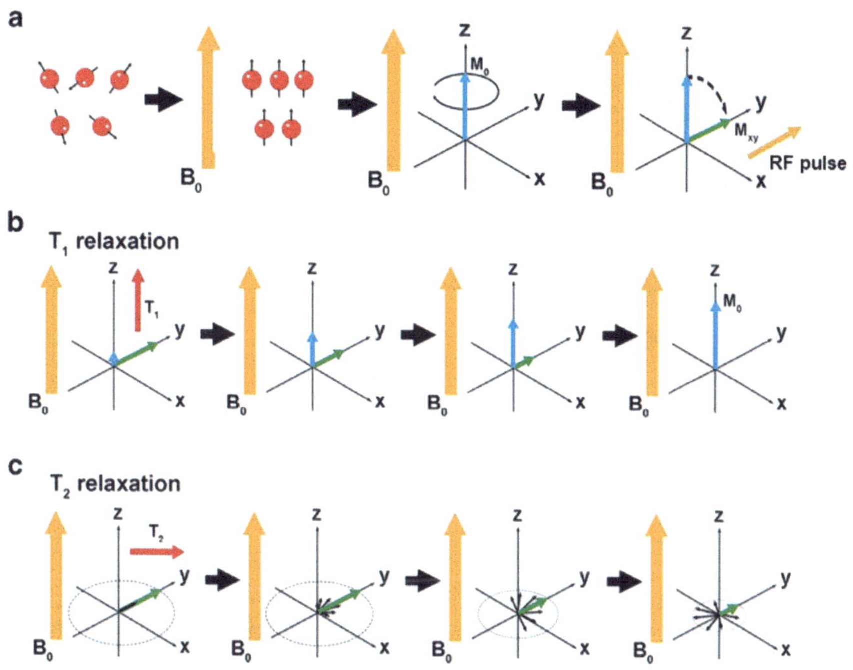

Fig. 9.5 Schematic representation of the basic principles of MRI. **a** When an external magnetic field B_0 (orange arrow) is applied (z-direction), protons (red spheres) tend to align with B_0 (note that the depiction here unrealistically shows all protons aligned with B_0, but there is actually only a slight preference for this alignment). This alignment results in a net magnetization vector (M_0, blue arrow). When an orthogonal RF pulse is applied, M_0 tilts 90° into the transverse (x–y) plane (M_{xy}, green arrow). The magnetization returns to equilibrium through two processes: T1 and T2 relaxation. **b** T1 relaxation: T1 is a measure of the time it takes the initial longitudinal magnetic moment (M_0) to recover. **c** T2 relaxation: T2 measures the loss of the transverse magnetic moment (M_{xy}) due to dephasing. Reproduced with permission from Jeon et al. (2021), Mastrogiacomo et al. (2019). Copyright (2019) American Chemical Society

coils to create variations in the primary magnetic field direction (Mastrogiacomo et al. 2019).

Current MRI contrast agents can be classified as superparamagnetic or paramagnetic based on their magnetic properties. Accordingly, two essential kinds of MRI contrast agents are soluble paramagnetic metal chelates and superparamagnetic iron oxide (SPION) nanoparticles (Thorek et al. 2006). Based on their biodistribution, the contrast agents can also be categorized as extracellular, blood pool (intravascular), and tissue or organ-specific (Yan et al. 2007). Iron oxide particles can be coated with a variety of macromolecular components and are made up of distinct iron oxide cores given in Fig. 9.7. The surfaces of the MNPs must be coupled with active targeting probes, including proteins and antibodies, in order to achieve more effective targeted imaging. These nanostructures can collect close to the targeted tissue because of the unique interactions between the targeting agent and receptors.

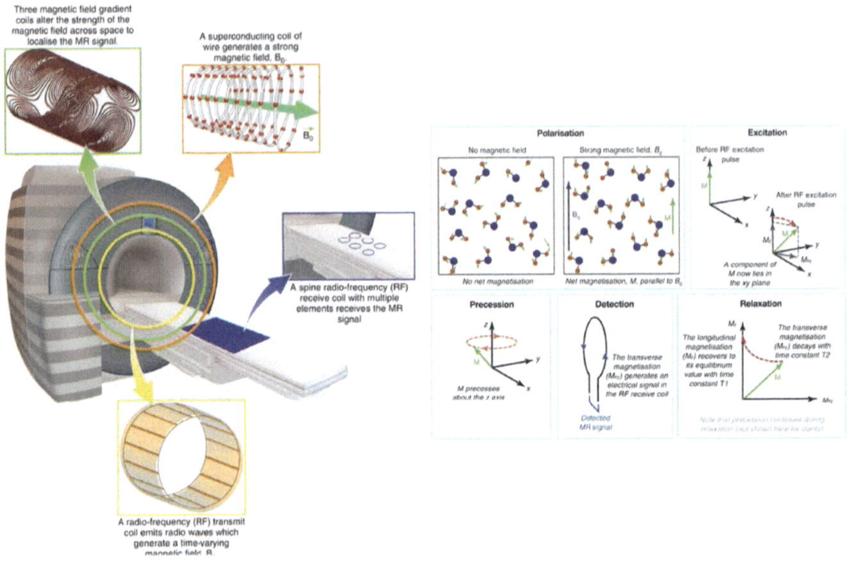

Fig. 9.6 Main components of an MRI scanner, direction of external magnetic field is in the head-foot direction in the scanner and processes involved in MRI. URL: *Magnetic resonance imaging (MRI) methods*

The transverse (T2) relaxation periods of water protons are to be changed by the T2 agent. Stem cells implanted in organs that show high signal intensity (such as kidney or lymphoid tissues) can be seen using T2 agents, which produce dark negative dark signal intensity in pictures.

If the magnetic ions in the cores are mutually aligned, they are considered to have superparamagnetic capabilities. These cores' composition and physicochemical characteristics alter constantly from several thousand magnetite Fe_3O_4 and $–Fe_2O_3$ crystals. Because of their significantly larger permanent magnetic moments when subjected to a magnetic field than a single paramagnetic molecule, these particles exhibit higher relaxivities than paramagnetic substances. SPION-based T2 agents seem to be the best MRI contrast agents for tracking stem cells because of their superior biocompatibility and high sensitivity when compared to T1 agents. Additionally, the size of the particles will affect how easily they can spread from the bloodstream into tumors. Designing stealthy nanoparticles requires consideration of both the particle size and the chemical structure of the surface. For example, the reticuloendothelial system (RES) interaction of SPIONs with a diameter $\leq$40 nm appears to be more affected by surface characteristics than size, according to some evidence presented by Rosen et al. in Table 9.2 (Rosen et al. 2012a, b) compiled from a study conducted by Qiao et al. (2009).

Noteworthy, Li et al. listed other magnetic-based MRI contrast agents in a very comprehensive table entitled "Some novel SPIONs as MRI contrast agents in stem cell labeling and tracking" in their study (Li et al. 2013).

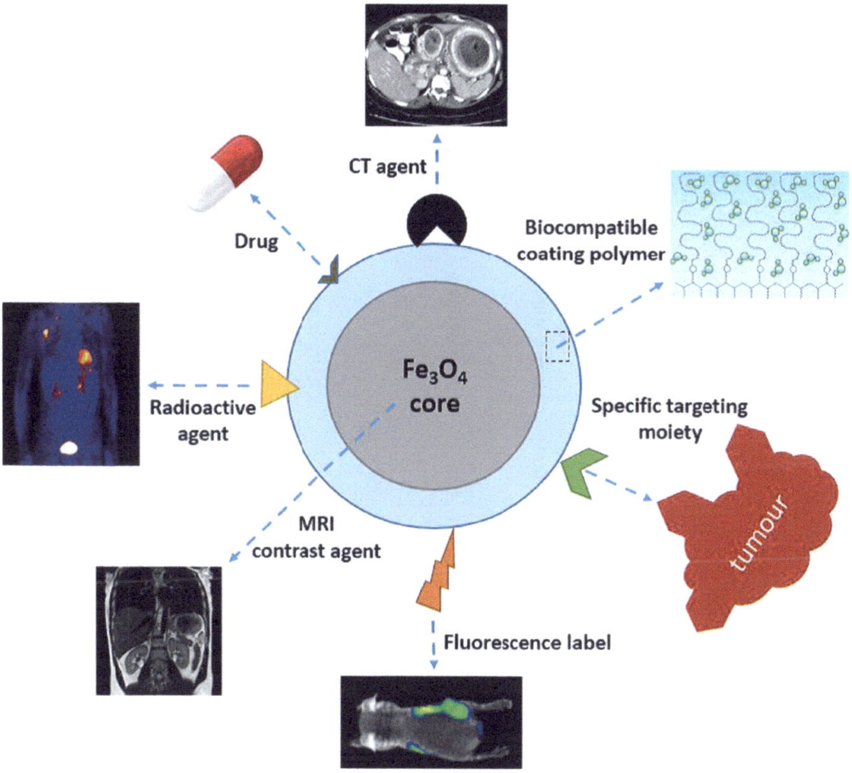

Fig. 9.7 Concept of a multimodal contrast agent based on a Fe_3O_4 core nanoplatform. Reproduced with permission from Clemente (2018)

SPIONs exhibit several advantages over traditional contrast agents used in MRI with high magnetic signal strength which enhances the contrast in MRI images. Longer lasting contrast enhancement can provide prolonged contrast enhancement in MRI scans compared to traditional agents, and low sensitivity to the number of water molecules unlike some traditional contrast agents, SPIONs do not rely on the number of water molecules surrounding them to generate contrast. These properties make SPIONs an attractive choice for molecular and cellular imaging, allowing for more accurate and sensitive detection of biological processes and structures (Bárcena et al. 2009). Additionally, investigations have shown that the iron released from degrading SPIONs is metabolized by the body, reducing the potential for long-term cytotoxicity considered nontoxic to cells and can be safely used in biomedical applications (Muthiah et al. 2013; Rosen et al. 2012a, b) (Table 9.3).

Typically, SPIONs have a way of distinguishing between healthy and malignant tissue. This facilitates their preferential accumulation in a specific tissue type, which

Table 9.2 SPION passive targeting-based contrast agents currently for sale or in clinical trials. Reproduced with permission (Rosen et al. 2012a, b)

Compound	Trade name	Hydro-dynamic size	Coating agent	Route of administration	Half-life (T1/2)	Indication	Development
Ferristene (OMP)	Abdoscan	3500 nm (composed of crystals < 50 nm)	Sulfonated styrene–divinylbenzene co-polymer	Oral	Oral	Gastrointestinal	Approval (Europe, discontinued)
Ferumoxsil (AMI-121)	GastroMARK Lumirem	300 nm (composed of 10 nm crystals)	Siloxane	Oral	IV bolus injection	2.4–3.6 h	Liver, spleen
Ferrixan (Ferucarbotran, SHU 555A)	Resovist Cliavist	60 nm	Carboxydextran	IV injection	6 min	Liver, spleen	Phase III (USA), approval (Japan; Europe; Australia)
Ferumoxide (AMI-25)	Feridex Endorem	120–180 nm; 80–150 nm	Dextran	IV bolus injection	6 h	Angiography	Approval (USA, 1996; Europe; Brazil; Japan; Argentina; South Korea; Israel; China; discontinued)
SHU 555 C	Supravist	≤20 nm	Dextran	IV bolus injection	6 h	Lymph node, liver, angiography	Phase I
Ferumoxtran-10 (AMI-227, BMS-180549)	Combidex Sineren	20–40 nm	Dextran	IV injection	24–36 h	Lymph node, liver, angiography	Phase III (discontinued)
Feruglose (PEG-feron, NC100150)	Clariscan	Core size 5–7 nm; total size 20 nm	Carbohydrate-polyethylene glycol coating (PEGylated starch)	IV injection	6 h	Lymph node, liver, perfusion, angiography	Preclinical (development discontinued)
(VSOP-C184)	–	7 nm	Citrate	IV injection	0.6–1.3 h	Angiography	Phase I

Table 9.3 Some commercial SPION-based MRI contrast agents ((Li et al. 2013)-stated that the table presented was extracted from (Laurent et al. 2008) and (Liu and Zhang 2012))

Brand name	Structure	Hydrodynamic size (nm)	Classification	Target
Ferumoxides (Feridex IV®, Endorem TM)	Dextran-coated SPIOs	80–150	T2 agent	Reticuloendothelial system Liver Stem cell labeling
Resovist	Carboxydextran-coated USPIOs	20	T2 agent	Blood pool Stem cell labeling
Sinerem	Dextran-coated USPIOs	15–30	T2 agent	Blood pool
Ferumoxytol®	Carboxymethyl-dextran coated USPIOs	30	T2 agent	Macrophage Blood pool
Ferumoxsil®	Silicon-coated SPIOs	300	T2 agent	Liver
Ferucarbotran® (SHU-555A)	Carboxydextran-coated USPIOs	60	T2 agent	Liver

lowers the T2 MRI signal and improves image contrast. SPIONs that are preferentially bound and absorbed by malignant tissues can be generated using a variety of techniques.

As Rosen et al. pointed out, targeting techniques typically fall into one of two categories: active targeting, which uses ligands conjugated to the surface of SPIONs to identify specific surface landmarks in cancerous tissue, or passive targeting, which depends on physiological differences between cancerous and normal tissues (Rosen et al. 2012a, b).

In a study Wang et al. reported that, nanoparticles exhibiting bright near-infrared (NIR) fluorescence, T1- and T2-weighted MR were synthesized for specific tumor imaging (Fig. 9.8). Clinically used Fe_3O_4 nanoparticles exhibit an intrinsic dark signal (T2-weighted MRI), which sometimes misleads clinical diagnosis (Wang et al. 2015). For both T1- and T2-weighted MRI and tumor fluorescence imaging, they combined ultra small Fe_3O_4 nanoparticles (2–3 nm) with a semiconducting polymer that emits infrared light. Folic acid-functionalized bio-functionalized multi-modality fluorescent magnetic nanoparticles (FMNPs) show high relaxation and strong fluorescence ($r_1 = 7.008$ m M^{-1} s^{-1}, $r_2 = 26.788$ m M^{-1} s^{-1}, $r_2/r_1 = 3.8$). These FMNPs have a lengthy circulation time, low aggregation, and a tiny average dynamic size of around 20 nm. According to in vitro research, FMNPs can be used as a fluorescent probe that targets human A549 lung cancer cells without clearly causing cytotoxicity. According to the results of in vivo experiments, FMNPs can preferentially accumulate in tumor tissues for T1- and T2-weighted MRI and other particular fluorescence imaging.

Noninvasive imaging techniques that differentiate molecular characteristics between cancerous and normal tissues hold significant promise for improving tumor detection. The transferrin receptor (TfR), which is often overexpressed in various cancers due to the malignant transformation of cells, represents a viable

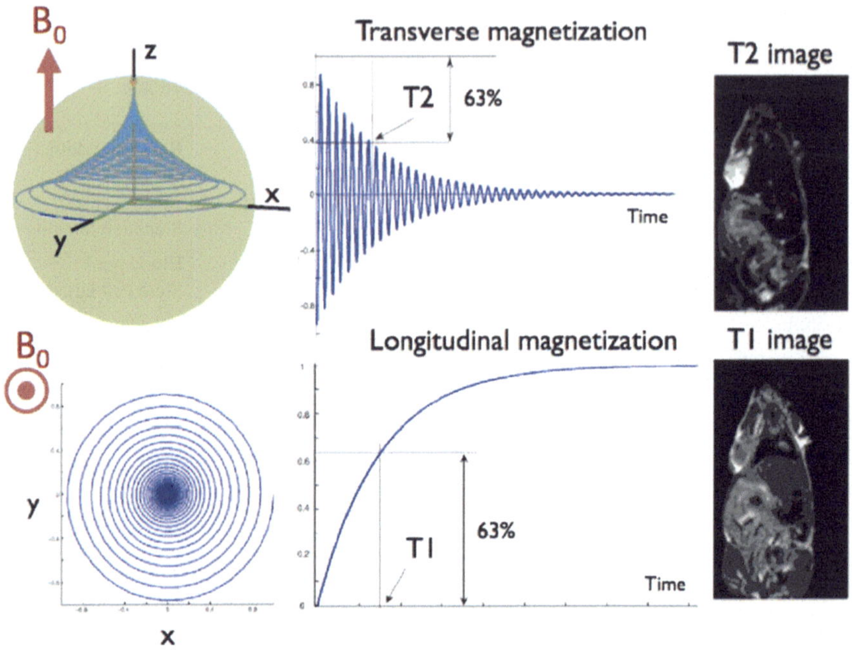

Fig. 9.8 T2 and T1 images in MRI. The relaxation time of transverse (T2) and longitudinal (T1) components of the magnetic moment to reach their equilibrium positions, aligned to B_0, are measured to produce T1 and T2 images. The images on the right display a coronal plane of a mouse. Reproduced with permission from Wang et al. (2015). Copyright (2015) The Royal Society of Chemistry

target for these imaging technologies. By focusing on TfR, it may be possible to enhance the identification of smaller tumors, thereby facilitating earlier diagnosis and intervention.

The study by Högemann-Savellano et al., for instance, investigates the in vivo testing of a new material, endogenous Tf-SPION conjugate in rats bearing 9L gliosarcoma tumors. Cross-linked iron oxide nanoparticles (CLIO) was treated with ammonia and obtained CLIO-NH_2, dialyzed to obtain reactive groups for the covalent binding of bifunctional linker reagents. The conjugation of CLIO-NH_2, they prepared the conjugation of CLIO-NH_2 and human diferric Tf using the linker reagent *N*-succinimidyl 3-(2-pyridyldithio)propionate leads to a disulfide bond (Tf-SS-CLIO), whereas employing the linker reagent succinimidyl iodoacetate forms a covalent linkage through a thioether bond (Tf-SC-CLIO). Their tests show that the new magnetic resonance imaging (MRI) probe is sensitive enough to identify minute variations in human cancer cell lines' endogenous TfR expression. Two tumors expressing either a high (9L3.9) or a low (9L4.2) level of the TfR were imaged before injection and 72 h after injection of Tf-SPION particles. The decrease in T2 relaxation time as seen in Fig. 9.9, was much more pronounced for the tumor

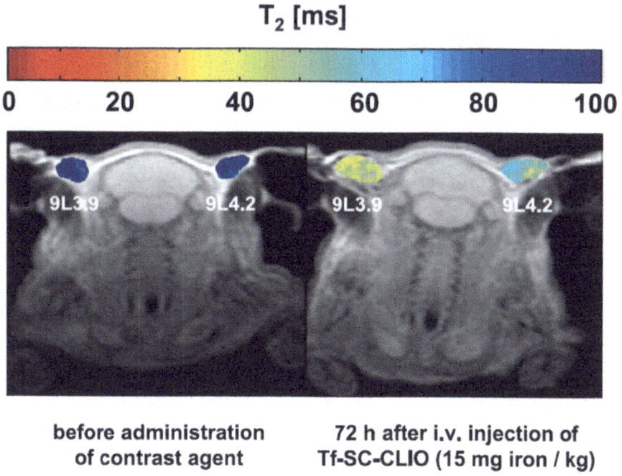

Fig. 9.9 Testing the Tf-SC-CLIO conjugate in vivo. Two cell lines expressing either a high (9L3.9 cells) or a low (9L4.2 cells) level of the TfR were implanted subcutaneously. Five days after implantation, MRI was performed (1.5 T, Signa 5.0, GE, 1.5-in. coil; spin echo sequence, TR 2000 ms, TE 15–10 ms) to generate a precontrast image. Following this precontrast MRI, Tf-SC-CLIO at a dose of 15 mg/kg iron was injected intravenously and MRI was repeated 24, 48, and 72 h later. The T2 values of the tumors of a representative animal superimposed with a proton-weighted image. Reproduced with permission from Högemann-Savellano et al. (2003). Copyright (2003) Elsevier

expressing high levels of the TfR compared to the tumor expressing low levels of the receptor (Högemann-Savellano et al. 2003).

Similarly, Cole et al. (2011) are interested in enhanced magnetic tumor targeting for imaging and designed new materials abbreviated as D5 (crosslinked) starch-coated MNP (modified with 5000 MW PEG) and D20 (modified with 20,000 MW PEG). Baseline T2-weighted, fast spin echo images indicate the position of the tumor (hyperintense region) in the brain. T2-weighted, gradient echo images provide qualitative information about MNP presence in the brain and tumor given in Fig. 9.10. Sustained negative contrast (hypointensity) in the tumor confirms that D5 (crosslinked) starch-coated MNP (modified with 5000 MW PEG) and D20 (modified with 20,000 MW PEG) reside for far longer in plasma and provide greater exposure to tumor compared to D (unmodified, parent starch MNP). Indeed, some hypointensity can still be observed in tumors through a period of 24 h for the PEGylated MNPs (Cole et al. 2011).

9.2.2 *Magnetofection*

A novel method to improve the delivery of gene vectors into cells is magnetofection. The genetic material can be delivered to the target cell surface using this technique. The DNA is released into the cytoplasm with the intention that the MNPs won't affect

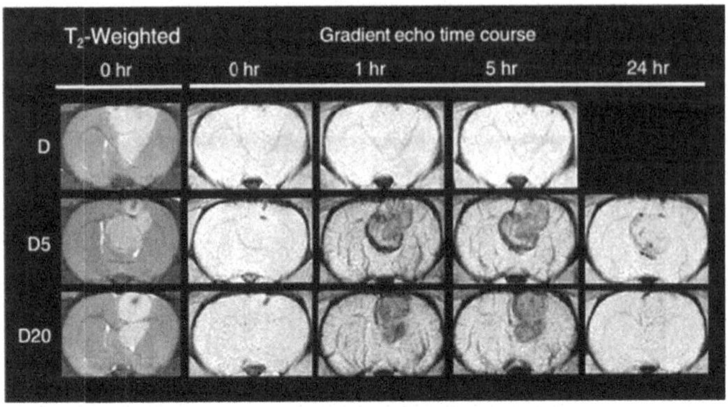

Fig. 9.10 Example of T2 magnetization relaxation time effect on cancer diagnosis Brain MRI time course of crosslinked starch-coated, PEG-modified MNPs in male Fisher 344 rats bearing 9L-glioma brain tumors (12 mg Fe/kg). Reproduced with permission from Cole et al. (2011). Copyright (2011) Elsevier

cellular activity (Scherer et al. 2002). When an external magnetic field is present, magnetic nanoparticles complex with the therapeutic gene are guided to the desired cells. High transfection efficiency has previously demonstrated that this method is a useful tool for both in vitro and in vivo gene transfer. Placing electromagnets beneath the cell culture apparatus creates an electric field gradient, which increases the penetration of the magnetic nanocomplexes into the cells and accounts for the technique's high transfection effectiveness (Sharma et al. 2021). These magnetic nanoparticles can be administered intravenously to produce effective in vivo transfection efficiency. The therapeutic genes are released by enzymatic cleavage, electrostatic interactions, or matrix degradation when the nanoparticles are kept at a target spot under the influence of high gradient electromagnetic fields (Bi et al. 2020).

Iron oxide particles are mostly used in magnetofection. Usually, these iron oxide particles are enclosed in a metallic or polymeric shell or distributed in a polymer matrix, such as dextran. SPIONs are commonly utilized in magnetofection due to their strong movement in response to external magnetic field gradients, with polyethyleneimine often used to form a shell around these particles to facilitate the electrostatic complexation with therapeutic genes. A key advantage of superparamagnetic particles is their low tendency to aggregate without an external magnetic field, enhancing their application in gene therapy. Magnetofection has proven effective for transfecting primary cells, which is typically challenging with traditional methods, and has been investigated for treating breast cancer and other solid tumors, as well as for intradermal gene therapy in ischemic skin flaps in rats. However, despite its promising results, magnetofection faces challenges such as low in vivo transfection efficiency, accumulation of iron oxide with repeated dosing, and rapid clearance from systemic circulation after intravenous administration. Additionally, effective use of this technique requires particles to be within a specific size range either less than 50 nm or larger than 5 μm (Sharma et al. 2021).

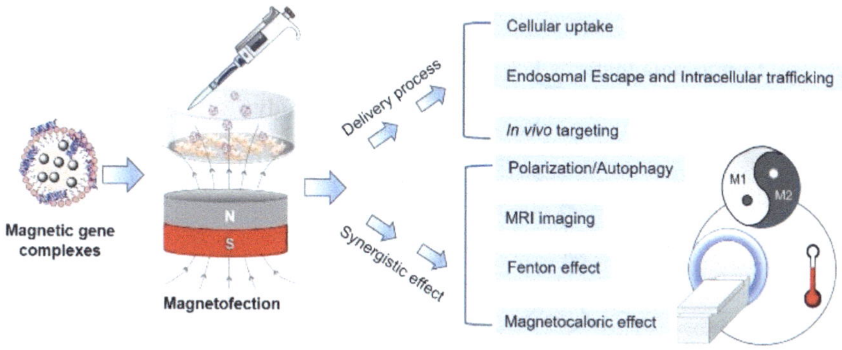

Fig. 9.11 Magnetic nanoparticles for efficient gene delivery and associated synergistic effects. Reproduced with permission from Bi et al. (2020). Copyright (2020) Elsevier

In recent years, the application of MNPs could promote the efficiency of gene transfection by holding the carriers at the targeting site via an external magnetic field. Bi et al. at their paper, focus on the application and mechanism of MNPs in gene delivery, as well as recent synergistic effects together with magnetofection (Bi et al. 2020). Magnetofection together with another synergistic effect for MRI with modified SPIONs could be used to deliver genes for diagnosis and treatment integration (Fig. 9.11). High gene transfection could simultaneously appear with MRI contrast, thus MRI guided visible gene transfection could be realized.

9.2.3 Magnetic Separation

Magnetically responsive nano- and microparticles and other relevant materials can be selectively separated (removed) from the complex samples using an external magnetic field (e.g., using an appropriate magnetic separator, permanent magnet, or electromagnet) (McCloskey et al. 2003). This process is very important for bioapplications since the absolute majority of biological materials have diamagnetic properties which enable efficient selective separation of magnetic materials. In parallel with the development of immunofluorescent markers, has also advanced including optical scanners and sorters which a new era in the development of separation methods and cell analysis. One significant aspect of magnetic cell separation is the degree to which a cell binds paramagnetic materials. The antigen–antibody reaction provides the potential for addressing the cell at a molecular level. Magnetic cell separation typically employs the use of antibodies against specific cell surface epitopes to tag cells of interest with magnetic particle conjugates in Fig. 9.12.

This cell separation method leverages the capability to label cell surface markers with antibodies tagged to magnetic beads, along with the ability of a magnetic field to attract these labeled particles from a distance. The controlled movement of these particles through magnetophoresis is essential for separating diverse cell populations

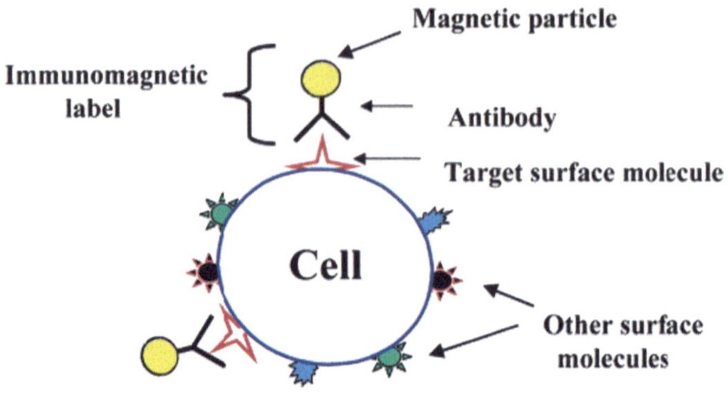

Fig. 9.12 Diagram of an immunomagnetically labeled cell. Note that the monoclonal antibodies are binding only to the specifically targeted cell surface molecules. Reproduced with permission from McCloskey et al. (2003). Copyright (2003) American Chemical Society

and forms the foundation of magnetic-activated cell sorting (MACS). Cells can be isolated using either tube-based or column-based techniques (Fig. 9.13) (Plouffe et al. 2015; Tomlinson et al. 2012).

High gradient magnetic cell sorting (MACS) has become increasingly popular as the method of choice for cell separation for various applications. It combines high sensitivity and high purity as well as increased recovery and viability of isolated cell populations compared to other methods. Magnetic cell sorting is particularly useful for isolation of rare cells from heterogeneous cell populations. The recent development of MACS eliminated the last drawback of magnetic cell sorting, namely the restriction to a single parameter. These new techniques add new parameters for

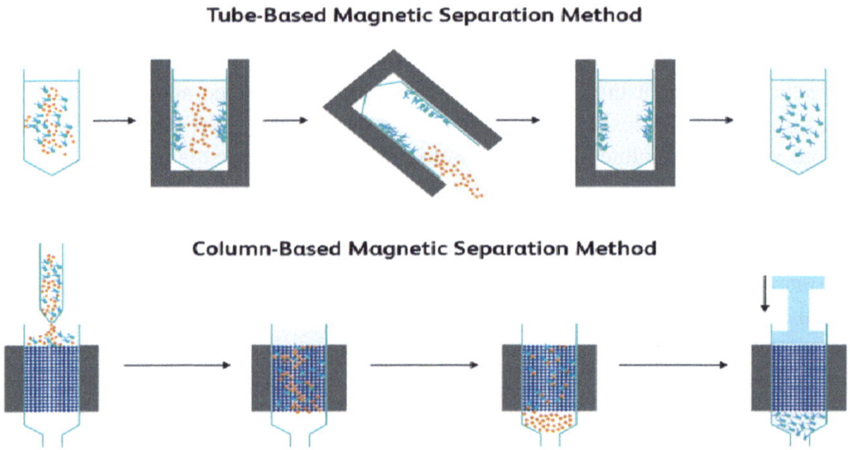

Fig. 9.13 Depiction of magnetic-activated cell sorting (MACS) by tube-based and column-based methods (BD Biosciences 2024)

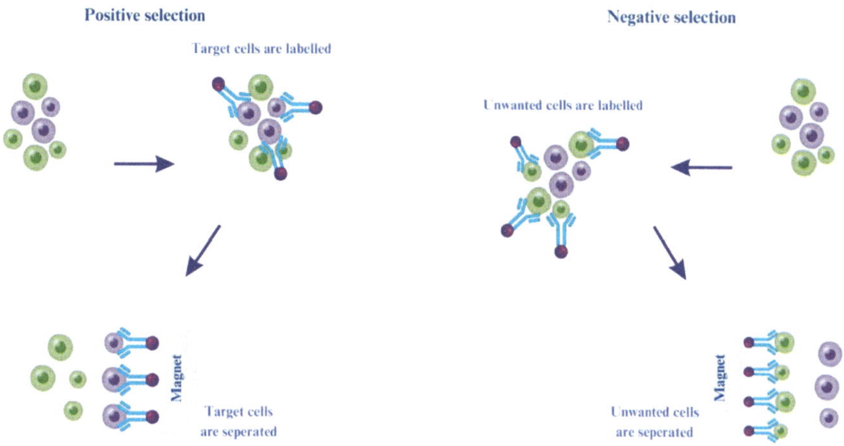

Fig. 9.14 Positive versus negative selection strategies

cytometric analysis and together with efficient magnetic separation, they provide new perspectives for research, as well as for the diagnosis and therapy of many diseases.

Magnetic cell separation also is a used method in recent years that allows for the rapid positive and/or negative selection of cell population in Fig. 9.14. Whether to label the target cells with antibodies for subsequent positive selection or to employ a negative selection strategy to eliminate an undesirable cell subset from the sample (unwanted cells are labeled) depends on the application. High purity is a positive selection strategy's primary benefit, and it works especially well for isolating rare, valuable cells. MACS also fluorescence-activated cell sorting (FACS) are among the most commonly used methods for cell separation, alongside density gradient techniques and adhesion-based approaches. FACS sorters are advanced devices that enable high cell sorting rates of over 50,000 cells per second and can achieve high purity levels close to 100% (Frenea-Robin and Marchalot 2022).

9.2.4 Cell Labeling

Magnetic labeling of living cells creates properties for many biomedical applications from individual cell manipulation to MRI tracking (Kolosnjaj-Tabi et al. 2013). It is a two-step process, involving the tagging or labeling of the desired biological entity with magnetic material, and the separating out of these tagged entities via a fluid-based magnetic separation device. To demonstrate the non-specific nature of labeling anionic magnetic nanoparticles (AMNPs) this labeling procedure was performed for a wide variety of mammalian cells, including phagocytic and non-phagocytic cells, different species (human, rat, mouse), different cell types (adult cells, progenitor or stem cells, immune cells and commonly used tumor cells), different cell size and culture properties (growth in confluence or in suspension, primary cultures or cell lineages (Wilhelm and Gazeau 2008). They compared the labeling mechanism, uptake biocompatibility, and efficiency with 14 different cell types. They also demonstrate that the same labels supply sufficient magnetization for MRI detection and distal manipulation. Labeling cells in vitro suggests the opportunity to control cell interaction with nanoparticles. Particles first adsorb on the plasma membrane, which consequently undergoes invagination. The MNP-loaded vesicles then pinch off the membrane and subsequently fuse with endosomes and lysosomes, which are dispersed within the cell's cytosol. While appropriate labeling conditions preserve the homeostasis of cells, failure to provide optimal labeling conditions might lead to particle aggregation that might have a negative impact on the cells.

To examine the effectiveness of magnetic labeling, Smirnov et al. investigated the viability of in vivo cellular imaging in mice using a 1.5 T clinical magnet. Anionic γ-Fe_2O_3 superparamagnetic iron oxide nanoparticles were used to mark hybridoma cells. The endocytose pathway internalized them (Smirnov et al. 2004). Particle uptake was measured using both electron spin resonance and magnetophoresis, which measures the migratory velocity of tagged cells in a magnetic field. With good agreement between the two techniques, it was discovered that hybridomas internalized nanoparticles quickly (less than two hours) and significantly (up to five pg of iron per cell). Male Swiss naked mice were given intraperitoneal injections of hybridomas tagged with 2.5 ng iron per cell. Additionally, a mean of 0.5 pg iron per cell—five times lower than that of the injected hybridomas—was discovered in labeled cells that were extracted from the spleen using ex vivo magnetic sorting in Fig. 9.15. This result is in line with these cells proliferating in vivo. Furthermore, it was discovered that the quantity of tagged hybridomas in the spleen was correlated with the strength of the MRI signal (Table 9.4).

9.3 Conclusion

Superparamagnetic nanoparticles (SPIONs) have emerged as pivotal tools in enhancing disease prevention, diagnosis, and treatment, particularly through their applications in magnetic resonance imaging (MRI), magnetofection, magnetic separation, and cell labeling.

MRI is a non-invasive imaging modality utilized by healthcare professionals for diagnosis and treatment planning. SPIONs exhibit significant potential as contrast agents in MRI due to their superparamagnetic properties, which enhance imaging quality. These nanoparticles can improve the visualization of tumors, making them valuable in cancer diagnostics. Their small size and biocompatibility allow for effective integration into biological systems, leading to advancements in imaging techniques that could revolutionize cancer detection and monitoring.

Magnetofection is another promising application of SPIONs, utilizing magnetic fields to facilitate the delivery of genetic material into cells. This method enhances transfection efficiency by concentrating the nanoparticles at the target site through an

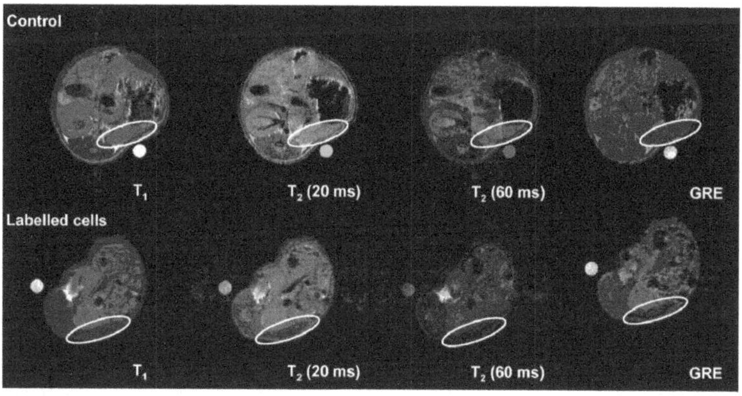

Fig. 9.15 In vivo cellular imaging of magnetically labeled hybridomas in the spleen with a 1.5 T clinical MRI system in vivo 1.5 T MRI. MRI of the spleen of a Swiss nude mouse 24 h after i.p. injection of magnetically labeled hybridomas (20 × 106) (bottom row), in comparison with uninjected control mice (top row). Sequence weightings (T1; GRE; T2 first and second echo, respectively) are indicated on the image. The spleen is indicated by an ellipse. A dark spleen signal was obtained relative to the control. Reprinted with permission from Smirnov et al. (2004). Copyright (2004) Wiley

external magnetic field, allowing for precise control over gene delivery. The ability of SPIONs to carry therapeutic genes or RNA molecules makes magnetofection a valuable strategy for gene therapy applications, potentially improving treatment outcomes for various diseases, including cancer.

Additionally, magnetic separation leverages the unique properties of SPIONs to isolate specific cell populations from heterogeneous mixtures rapidly and efficiently. This technique is particularly useful in biomedical research and clinical diagnostics, where the purification of target cells is essential. The specificity and speed of magnetic separation methods provide significant advantages over traditional techniques, enabling researchers to obtain high-purity samples with minimal processing time.

Cell labeling with SPIONs allows for the identification and tracking of specific cell populations in vivo. This technique is instrumental in studying cell migration and behavior in various medical assays. By tagging cells with iron oxide nanoparticles, researchers can monitor cellular dynamics using MRI or other imaging modalities. The labeling process is characterized by its simplicity and efficiency, making it applicable across diverse biomedical fields.

The integration of MRI with magnetic cell labeling is gaining traction as a preferred method for monitoring cell migration in medical assays. Iron oxide nanoparticles serve as effective agents for labeling various cell types, facilitating a range of biomedical applications from basic research to clinical diagnostics. The use of anionic nanoparticles for cell labeling presents several advantages, including ease of use, rapid application, and predictable efficiency. In summary, SPIONs are at the forefront of innovative biomedical applications, offering enhanced capabilities

Table 9.4 Cell types that have been labeled with AMNPs. Reprinted with permission from Wilhelm and Gazeau (2008)

	d (mm)	*K* $(mM)^{-1}$	m_0 (pg)	m_i (h)		m_p (pg)	Refs.
Immune cells							
RefRaw macrophages (mouse)	11.7 ± 0.8	17 ± 4	6 ± 0.2	1.3 ± 0.2	25 ± 5	33	Wilhelm et al. (2003)
Hybridomas (mouse)	12	44 ± 31	2.4 ± 0.2	0.4 ± 0.1	1.5 ± 0.3	5.5	Smirnov et al. (2004)
Dendritic cells (human)	12.2 ± 1.6	29 ± 8	6.6 ± 0.4	0.9 ± 0.1	1.8 ± 0.5	15.2	Wilhelm et al. (2002a, b)
OT-1 lymphocytes (mouse)	8.4 ± 0.6	43 ± 26	1.3 ± 0.1	2 ± 0.8	1.1 ± 0.5	22	Smirnov et al. (2006)
EL4-B lymphocytes (human)	9.2 ± 1.5	30 ± 10	4.1 ± 0.2	1.4 ± 0.2	1.3 ± 0.4	8	Wilhelm and Gazeau (2008)
Tumor cells							
HeLa ovarian carcinoma (human)	20.2 ± 2.6	17 ± S	18 ± 0.3	1 ± 0.1	2.5 ± 0.2	37.5	Wilhelm et al. (2002a, b)
PC3 prostatic carcinoma (human)	14.6 ± 2.3	16 ± S	8.1 ± 0.2	0.4 ± 0.1	1.8 ± 0.3	13.4	Fortin et al. (2008), Wilhelm et al. (2007b)
HuH7 hepatic carcinoma (human)	11.6 ± 1.7	27 ± 11	3.4 ± 0.2	1.2 ± 0.1	1.5 ± 0.2	8.3	Wilhelm and Gazeau (2008)
Therapeutic adult cells							
Hepatocytes (mouse)	20 ± 4	30 ± 13	21.3 ± 0.8	1.4 ± 0.2	1.9 ± 0.3	49	Luciani et al. (2008)
Gingival fibroblasts (human)	17 ± 3	24 ± 8	13.5 ± 0.6	1.6 ± 0.5	1.6 ± 1	28	Wilhelm et al. (2007a, b)
Smooth muscle cells (rat)	14 ± 1	29 ± 8	S.8 ± 0.2	1.5 ± 0.1	1.9 ± 3	12.4	Rivière et al. (2005)
Therapeutic stem cells or progenitor cells							

(continued)

Table 9.4 (continued)

	d (mm)	*K* (mM)$^{-1}$	m_0 (pg)	m_i (h)		m_p (pg)	Refs.
Myogenic precursor cells (pig)	14 ± 2.5	22 ± 11	3.5 ± 0.2	1.8 ± 0.3	1.1 ± 0.3	48	Rivière et al. (2009)
Endothelial progenitor cells (human)	13.8 ± 2.1	39 ± 12	8.9 ± 0.3	1.4 ± 0.2	1.9 ± 0.3	20.5	Wilhelm et al. (2007a, b)

d is the cell diameter, K is the affinity constant of AMNPs for the cell membrane, m_0 is the binding capacity on plasma membrane (in mass of attached particles), m_i (h) is the characteristic time for internalization, is the maximal fraction of internalized membrane, m_p is the predicted mass of iron per cell for a labeling condition of [Fe] = 20 mM for 2 h at 37 °C

in disease diagnosis and treatment strategies through their multifunctional properties across MRI, magnetofection, magnetic separation, and cell labeling.

Abbreviations

NP	Nanoparticle
MNP	Magnetic nanoparticle
MRI	Magnetic resonance imaging
SPIONs	Superparamagnetic iron oxide nanoparticles
CNT	Carbon nanotube
MACS	Magnetic cell sorting
RF	Radiofrequency
CLIO	Cross-linked iron oxide nanoparticles
NIR	Near-infrared
TfR	Transferrin receptor
RES	Reticuloendothelial system
USPIO	Ultra-small superparamagnetic iron oxide
FMNPs	Fluorescent magnetic nanoparticles
MACS	Magnetic-activated cell sorting
FACS	Fluorescent-activated cell sorting
AMNPs	Anionic magnetic nanoparticles
EPR	Enhanced permeability and retention
WHO	World Health Organization
PET	Positron emission tomography
PEG	Polyethylene glycol
SPECT	Single photon emission computed tomography
CT	Computed tomography

References

Bárcena C, Sra AK, Gao JM (2009) Applications of magnetic nanoparticles in biomedicine. In: Nanoscale magnetic materials and applications, pp 591–626. https://doi.org/10.1007/978-0-387-85600-1_20

Bardhan R, Lal S, Joshi A, Halas NJ (2011) Theranostic nanoshells: from probe design to imaging and treatment of cancer. Acc Chem Res 44(10):936–946. https://doi.org/10.1021/ar200023x

BD Biosciences (2024, seen December 28). https://www.bdbiosciences.com/en-us/learn/applications/magnetic-separation?tab=Overview

Belyanina I, Kolovskaya O, Zamay S, Gargaun A, Zamay T, Kichkailo A (2017) Targeted magnetic nanotheranostics of cancer. Molecules 22(6). https://doi.org/10.3390/molecules22060975

Bi Q, Song X, Hu A, Luo T, Jin R, Ai H, Nie Y (2020) Magnetofection: magic magnetic nanoparticles for efficient gene delivery. Chin Chem Lett 31(12):3041–3046. https://doi.org/10.1016/j.cclet.2020.07.030

Clemente GS (2018) Radiolabelling of nanomaterials for positron emission tomography imaging. Master of Science (MSc) in biological sciences, The University of Hull

Cole AJ, David AE, Wang J, Galbán CJ, Hill HL, Yang VC (2011) Polyethylene glycol modified, cross-linked starch-coated iron oxide nanoparticles for enhanced magnetic tumor targeting. Biomaterials 32(8):2183–2193. https://doi.org/10.1016/j.biomaterials.2010.11.040

Coskun S (2016) The applications of magnetic nanoparticles in cancer diagnosis. Graduation thesis, Bezmialem Vakif University, Istanbul

Currie S, Hoggard N, Craven IJ, Hadjivassiliou M, Wilkinson ID (2013) Understanding MRI: basic MR physics for physicians. Postgrad Med J 89(1050):209–223. https://doi.org/10.1136/postgradmedj-2012-131342

Davis ME, Chen Z, Shin DM (2008) Nanoparticle therapeutics: an emerging treatment modality for cancer. Nat Rev Drug Discovery 7(9):771–782. https://doi.org/10.1038/nrd2614

Dobrovolskaia MA, Mcneil SE (2007) Immunological properties of engineered nanomaterials. Nat Nanotechnol 2(8):469–478. https://doi.org/10.1038/nnano.2007.223

Ferrari M (2005) Cancer nanotechnology: opportunities and challenges. Nat Rev Cancer 5(3):161–171. https://doi.org/10.1038/nrc1566

Fortin J-P, Gazeau F, Wilhelm C (2008) Intracellular heating of living cells through Néel relaxation of magnetic nanoparticles. Eur Biophys J 37(2):223–228. https://doi.org/10.1007/s00249-007-0197-4

Frenea-Robin M, Marchalot J (2022) Basic principles and recent advances in magnetic cell separation. Magnetochemistry 8(1). https://doi.org/10.3390/magnetochemistry8010011

Gao XH, Cui YY, Levenson RM, Chung LWK, Nie SM (2004) Cancer targeting and imaging with semiconductor quantum dots. Nat Biotechnol 22(8):969–976. https://doi.org/10.1038/nbt994

Gao JH, Gu HW, Xu B (2009) Multifunctional magnetic nanoparticles: design, synthesis, and biomedical applications. Acc Chem Res 42(8):1097–1107. https://doi.org/10.1021/ar9000026

He CB, Liu DM, Lin WB (2015) Nanomedicine applications of hybrid nanomaterials built from metal-ligand coordination bonds: nanoscale metal-organic frameworks and nanoscale coordination polymers. Chem Rev 115(19):11079–11108. https://doi.org/10.1021/acs.chemrev.5b00125

Henglein A (1989) Small-particle research—physicochemical properties of extremely small colloidal metal and semiconductor particles. Chem Rev 89(8):1861–1873. https://doi.org/10.1021/cr00098a010

Högemann-Savellano D, Bos E, Blondet C, Sato F, Abe T, Josephson L, Weissleder R, Gaudet J, Sgroi D, Peters PJ, Basilion JP (2003) The transferrin receptor: a potential molecular imaging marker for human cancer. Neoplasia 5(6):495–506. https://doi.org/10.1016/S1476-5586(03)80034-9

Hosu O, Tertis M, Cristea C (2019) Implication of magnetic nanoparticles in cancer detection, screening and treatment. Magnetochemistry 5(4). https://doi.org/10.3390/magnetochemistry5040055

Huang XH, El-Sayed IH, Qian W, El-Sayed MA (2006) Cancer cell imaging and photothermal therapy in the near-infrared region by using gold nanorods. J Am Chem Soc 128(6):2115–2120. https://doi.org/10.1021/ja057254a
Jeon M, Halbert MV, Stephen ZR, Zhang MA-O (2021) Iron oxide nanoparticles as T(1) contrast agents for magnetic resonance imaging: fundamentals, challenges, applications, and prospectives, vol 33, no 1906539. https://doi.org/10.1002/adma.201906539
Kim CS, Duncan B, Creran B, Rotello VM (2013) Triggered nanoparticles as therapeutics. Nano Today 8(4):439–447. https://doi.org/10.1016/j.nantod.2013.07.004
Kolosnjaj-Tabi J, Wilhelm C, Clément O, Gazeau F (2013) Cell labeling with magnetic nanoparticles: opportunity for magnetic cell imaging and cell manipulation. J Nanobiotechnol 11(1):S7. https://doi.org/10.1186/1477-3155-11-S1-S7
Laurent S, Forge D, Port M, Roch A, Robic C, Vander Elst L, Muller RN (2008) Magnetic iron oxide nanoparticles: synthesis, stabilization, vectorization, physicochemical characterizations, and biological applications. Chem Rev 108(6):2064–2110. https://doi.org/10.1021/cr068445e
Li L, Jiang W, Luo K, Song H, Lan F, Wu Y, Gu Z (2013) Superparamagnetic iron oxide nanoparticles as MRI contrast agents for non-invasive stem cell labeling and tracking [review]. Theranostics 3(8):595–615. https://doi.org/10.7150/thno.5366
Liu Y, Zhang N (2012) Gadolinium loaded nanoparticles in theranostic magnetic resonance imaging. Biomaterials 33(21):5363–5375. https://doi.org/10.1016/j.biomaterials.2012.03.084
Luciani A, Parouchev A, Smirnov P, Braga G, Wilhelm C, Gazeau F, Boudechiche L, L'Hermine-Coulomb A, Dagher I, Franco D, Rahmouni A, Hadchouel M, Weber A, Clement O (2008) In vivo imaging of transplanted hepatocytes with a 1.5-T clinical MRI system—initial experience in mice. Eur Radiol 18(1):59–69. https://doi.org/10.1007/s00330-007-0750-7
Magnetic resonance imaging (MRI) methods. https://neupsykey.com/magnetic-resonance-imaging-mri-methods/
Mahmoudi M, Sant S, Wang B, Laurent S, Sen T (2011) Superparamagnetic iron oxide nanoparticles (SPIONs): development, surface modification and applications in chemotherapy. Adv Drug Deliv Rev 63(1–2):24–46. https://doi.org/10.1016/j.addr.2010.05.006
Mastrogiacomo SA-O, Dou W, Jansen JA, Walboomers XF (2019) Magnetic resonance imaging of hard tissues and hard tissue engineered bio-substitutes. Mol Imaging Biol (21):1003–1019. https://doi.org/10.1007/s11307-019-01345-2
McCloskey KE, Chalmers JJ, Zborowski M (2003) Magnetic cell separation: characterization of magnetophoretic mobility. Anal Chem 75(24):6868–6874. https://doi.org/10.1021/ac034315j
Mura S, Nicolas J, Couvreur P (2013) Stimuli-responsive nanocarriers for drug delivery. Nat Mater 12(11):991–1003. https://doi.org/10.1038/Nmat3776
Muthiah M, Park IK, Cho CS (2013) Surface modification of iron oxide nanoparticles by biocompatible polymers for tissue imaging and targeting. Biotechnol Adv 31(8):1224–1236. https://doi.org/10.1016/j.biotechadv.2013.03.005
Petros RA, DeSimone JM (2010) Strategies in the design of nanoparticles for therapeutic applications. Nat Rev Drug Discovery 9(8):615–627. https://doi.org/10.1038/nrd2591
Plewes DB, Kucharczyk W (2012) Physics of MRI: a primer (1522–2586 (Electronic))
Plouffe BD, Murthy SK, Lewis LH (2015) Fundamentals and application of magnetic particles in cell isolation and enrichment: a review. Rep Prog Phys 78(1):016601. https://doi.org/10.1088/0034-4885/78/1/016601
Qiao R, Yang C, Gao M (2009) Superparamagnetic iron oxide nanoparticles: from preparations to in vivo MRI applications. J Mater Chem 19(35):6274–6293. https://doi.org/10.1039/B902394A
Reisz JA, Bansal N, Qian J, Zhao W, Furdui CM (2014) Effects of ionizing radiation on biological molecules—mechanisms of damage and emerging methods of detection. Antioxid Redox Signal 21(2):260–292. https://doi.org/10.1089/ars.2013.5489
Rivière C, Boudghène FP, Gazeau F, Roger J, Pons J-N, Laissy J-P, Allaire E, Michel J-B, Letourneur D, Deux J-F (2005) Iron oxide nanoparticle-labeled rat smooth muscle cells: cardiac MR imaging for cell graft monitoring and quantitation. Radiology 235(3):959–967. https://doi.org/10.1148/radiol.2353032057

Rivière C, Lecoeur C, Wilhelm C, Péchoux C, Combrisson H, Yiou R, Gazeau F (2009) The MRI assessment of intraurethrally—delivered muscle precursor cells using anionic magnetic nanoparticles. Biomaterials 30(36):6920–6928. https://doi.org/10.1016/j.biomaterials.2009.08.056
Roeder RK (2013) Mechanical characterization of biomaterials. In: Characterization of biomaterials, pp 49–104. https://doi.org/10.1016/B978-0-12-415800-9.00003-6
Rosen JE, Chan L, Shieh D-B, Gu FX (2012a) Iron oxide nanoparticles for targeted cancer imaging and diagnostics. Nanomedicine: Nanotechnol, Biol Med 8(3):275–290. https://doi.org/10.1016/j.nano.2011.08.017
Rosen JE, Chan L, Shieh DB, Gu FX (2012b) Iron oxide nanoparticles for targeted cancer imaging and diagnostics. Nanomedicine: Nanotechnol Biol Med 8(3):275–290. https://doi.org/10.1016/j.nano.2011.08.017
Scherer F, Anton M, Schillinger U, Henke J, Bergemann C, Krüger A, Gänsbacher B, Plank C (2002) Magnetofection: enhancing and targeting gene delivery by magnetic force in vitro and in vivo. Gene Ther 9(2):102–109. https://doi.org/10.13140/RG.2.2.34195.14883
Shah FD, Begum R, Vajaria BN, Patel KR, Patel JB, Shukla SN, Patel PS (2011) A review on salivary genomics and proteomics biomarkers in oral cancer. Indian J Clin Biochem 26(4):326–334. https://doi.org/10.1007/s12291-011-0149-8
Sharma D, Arora S, Singh J, Layek B (2021) A review of the tortuous path of nonviral gene delivery and recent progress. Int J Biol Macromol 183:2055–2073. https://doi.org/10.1016/j.ijbiomac.2021.05.192
Shen Z, Wu A, Chen X (2017) Iron oxide nanoparticle based contrast agents for magnetic resonance imaging. Mol Pharm 14(5):1352–1364. https://doi.org/10.1021/acs.molpharmaceut.6b00839
Smirnov P, Gazeau F, Lewin M, Bacri JC, Siauve N, Vayssettes C, Cuénod CA, Clément O (2004) In vivo cellular imaging of magnetically labeled hybridomas in the spleen with a 1.5-T clinical MRI system. Magn Reson Med 52(1):73–79. https://doi.org/10.1002/mrm.20121
Smirnov P, Lavergne E, Gazeau F, Lewin M, Boissonnas A, Doan B-T, Gillet B, Combadière C, Combadière B, Clément O (2006) In vivo cellular imaging of lymphocyte trafficking by MRI: a tumor model approach to cell-based anticancer therapy. Magn Reson Med 56(3):498–508. https://doi.org/10.1002/mrm.20996
Thakur P, Taneja S, Chahar D, Ravelo B, Thakur A (2021) Recent advances on synthesis, characterization and high frequency applications of Ni-Zn ferrite nanoparticles. J Magn Magn Mater 530. https://doi.org/10.1016/j.jmmm.2021.167925
Thorek DLJ, Chen AK, Czupryna J, Tsourkas A (2006) Superparamagnetic iron oxide nanoparticle probes for molecular imaging. Ann Biomed Eng 34(1):23–38. https://doi.org/10.1007/s10439-005-9002-7
Tian J, Zhong J (2013) Molecular imaging: fundamentals and applications. In: Tian J (ed) Molecular imaging: fundamentals and applications. Springer Berlin Heidelberg, pp 3–14. https://doi.org/10.1007/978-3-642-34303-2_1
Tomlinson MJ, Tomlinson S, Yang XB, Kirkham J (2012) Cell separation: terminology and practical considerations. J Tissue Eng 4:2041731412472690. https://doi.org/10.1177/2041731412472690
Tran N, Webster TJ (2010) Magnetic nanoparticles: biomedical applications and challenges. J Mater Chem 20(40):8760–8767. https://doi.org/10.1039/c0jm00994f
Wang G, Zhang X, Liu Y, Hu Z, Mei X, Uvdal K (2015) Magneto-fluorescent nanoparticles with high-intensity NIR emission, T1- and T2-weighted MR for multimodal specific tumor imaging. J Mater Chem B 3(15):3072–3080. https://doi.org/10.1039/C5TB00155B
Wilhelm C, Gazeau F, Bacri JC (2002a) Magnetophoresis and ferromagnetic resonance of magnetically labeled cells. Eur Biophys J 31(2):118–125. https://doi.org/10.1007/s00249-001-0200-4
Wilhelm C, Gazeau F, Roger J, Pons JN, Bacri JC (2002b) Interaction of anionic superparamagnetic nanoparticles with cells: kinetic analyses of membrane adsorption and subsequent internalization. Langmuir 18(21):8148–8155. https://doi.org/10.1021/la0257337

Wilhelm C, Gazeau F (2008) Universal cell labelling with anionic magnetic nanoparticles. Biomaterials 29(22):3161–3174. https://doi.org/10.1016/j.biomaterials.2008.04.016

Wilhelm C, Billotey C, Roger J, Pons JN, Bacri JC, Gazeau F (2003) Intracellular uptake of anionic superparamagnetic nanoparticles as a function of their surface coating. Biomaterials 24(6):1001–1011. https://doi.org/10.1016/S0142-9612(02)00440-4

Wilhelm C, Bal L, Smirnov P, Galy-Fauroux I, Clément O, Gazeau F, Emmerich J (2007a) Magnetic control of vascular network formation with magnetically labeled endothelial progenitor cells. Biomaterials 28(26):3797–3806. https://doi.org/10.1016/j.biomaterials.2007.04.047

Wilhelm C, Fortin JP, Gazeau F (2007b) Tumor cell toxicity of intracellular hyperthermia mediated by magnetic nanoparticles. (1533–4880 (Print)). https://doi.org/10.1166/jnn.2007.668

Wilhelm S, Tavares AJ, Dai Q, Ohta S, Audet J, Dvorak HF, Chan WCW (2016) Analysis of nanoparticle delivery to tumors. Nat Rev Mater 1(5). https://doi.org/10.1038/natrevmats.2016.14

Yan G-P, Robinson L, Hogg P (2007) Magnetic resonance imaging contrast agents: overview and perspectives. Radiography 13:e5–e19. https://doi.org/10.1016/j.radi.2006.07.005

Zhao ZH, Zhou ZJ, Bao JF, Wang ZY, Hu J, Chi XQ, Ni KY, Wang RF, Chen XY, Chen Z, Gao JH (2013) Octapod iron oxide nanoparticles as high-performance contrast agents for magnetic resonance imaging. Nat Commun 4. https://doi.org/10.1038/ncomms3266

Chapter 10
Significance of Life Cycle Assessment of Clean Wastewater Treatment Technologies using Polymers to Achieve Sustainability

Neha Patni, Vruti Shah, Priya Shah, and Himandri Zala

Abstract Although water covers around 71% of the earth's surface, 98% of that water is unusable due to its salinity and cannot be utilised as drinking water. Two of the most important problems facing humanity in the twenty-first century are the scarcity of water and the quality of drinking water. There are several reasons why water can become contaminated, but chemical pollution poses a serious worldwide risk due to its extreme toxicity. A life cycle assessment (LCA) is a standardised method to quantitatively assess such environmental impacts. This study goes beyond LCA of wastewater treatment technologies for the sustainable evolution of enterprises in order to address these concerns. Hotspots are those points in the life cycle that have substantial negative influence on the environment. Possible hotspots can be determined using it, trade-offs can be comprehended, and more environmentally friendly wastewater treatment technologies can be created. Most often, resolving hotspots becomes the cornerstone of the sustainability plan based on a complete LCA. Existence Cycle Assessment (LCA) is a thorough methodology that evaluates how different products, activities, or systems affect the environment over their whole life. Conceptual, simplified and detailed LCA is utmost important to evaluate the sustainability of wastewater treatment systems. This chapter walks the reader through the LCA technique step-by-step so they may decide on the scope and aim with knowledge. It also includes variables to include and their sources for inventory acquisition, as well as the selection of the functional unit and system boundaries. Viewpoints on impact assessment method selection and effective strategy selection are also covered.

Keywords Life cycle assessment · Sustainability · Waste water · Water treatment · Environment

N. Patni (✉) · V. Shah · P. Shah · H. Zala
Chemical Engineering Department, Institute of Technology, Nirma University, Ahmedabad, Gujarat, India
e-mail: neha.patni@nirmauni.ac.in

P. M. Visakh (ed.), *Engineering Applications of Polymer-Based Nanocomposites*, Advanced Structured Materials 237, https://doi.org/10.1007/978-3-031-93463-6_10

10.1 Introduction

In today's world when environmental problems are felt on a global scale, wastewater treatment plants (WWTPs) play a crucial role in protecting ecosystems and ensuring that future generations have access to clean water. Such essential facilities are responsible for the large-scale processing and purification of wastewater, the removal of pollutants, and the mitigation of potential harm to human health and aquatic life. However, in order to achieve sustainable growth, it is essential to consider not just the immediate environmental impacts but also the entire life cycle of these plants. Investing in wastewater treatment operations and infrastructure is an important part of managing the quality of discharged wastewater, but so is ensuring that pollutant concentrations remain acceptable and satisfy contamination threshold values (Abily et al. 2023). Water treatment plants that take activated sludge quickly developed from their original use more than a century ago. Nevertheless, despite their extensive historical background and use, conventional activated sludge (CAS) facilities have transformed into an impractical method of treating wastewater. Furthermore, conventional wastewater treatment plants are tremendously energy-intensive and, most likely, allow for limited material retrieval (Guven et al. 2023). Therefore, by treating wastewater and lowering the discharge of toxins into the environment, wastewater treatment facilities significantly contribute to the reduction of environmental pollution. Nevertheless, their activities might nevertheless leave an ecological impact. So, to measure the environmental effect of plants in many ways, Life Cycle Assessment (LCA) offers a thorough framework for wastewater treatment (Tsangas et al. 2023). One such standardised and advanced method for evaluating the inputs, outputs, and possible environmental consequences of a product system across its lifetime is life cycle assessment (LCA).

Life Cycle Assessment (LCA) provides a thorough procedure for assessing WWTPs' environmental performance overall (Hellweg et al. 2023). With the use of life cycle assessment, we can assess the environmental costs and advantages related to the abstraction of raw materials, building, operation, maintenance, and eventually decommissioning of wastewater treatment plants. Through Life Cycle Assessment, we can determine possible hotspots, comprehend trade-offs, and create more environmentally friendly wastewater treatment technologies. Hotspots are those points in the life cycle that have substantial negative influence on the environment. Most often, resolving hotspots becomes the cornerstone of the sustainability plan based on a complete LCA. One useful method for determining how wastewater treatment facilities (WWTPs) affect the environment is LCA. Nevertheless, there has been little success thus far in implementing LCA methodologies into wastewater treatment systems. To help decision-makers find the best eco-friendly solutions, LCA studies analyse the environmental implications of various scenarios or treatment technologies. With this data, the plant can better allocate resources, run more efficiently, and reduce their impact on the environment. Hence in this chapter first the types of pollution are elaborated, followed by the basics of LCA. Moreover, different approaches to pursue it are discussed followed by case studies to support the statements.

10.2 Pollution of Water

The detrimental effects of water contamination on human health have prompted significant concerns about the issue. Ensuring that everyone has access to clean water, enhancing public health, and halting the development of water-related ailments are all vital for preserving life. But there are a number of pollutants that can lower water quality to an unsafe level, rendering it unsuitable for human consumption and everyday usage. Diseases caused by microbes, heavy metals, chemical pollutants, pesticides, medications, and novel toxins are all potential water contaminants. Chemical contaminants, such as common household items, agricultural runoff, and industrial waste, pose a threat to human health because they contaminate water supplies. Ongoing research is being conducted to determine the effects of new pesticides, medications, and emerging pollutants on human health (Babuji et al. 2023). Water sources such as rivers, lakes, seas, groundwater, and even smaller volumes of water like ponds or streams, can get contaminated or contain hazardous substances. These pollutants have a range of detrimental effects on the surrounding ecology, human health, and aquatic life when they enter the water either directly or indirectly. That can be further classified into different types as discussed further.

10.2.1 Point Source Pollution

Contamination that originates from a single point is known as point source pollution. Effluent, or wastewater, is one example of a substance that might be legally or illegally emitted by a producer, an oil refinery, or a wastewater treatment plant. Unlawful dumping, chemical and oil spills, and leaky septic systems are further sources of pollution. Point source pollution is managed by the EPA by limiting the types of establishments that are authorised to discharge directly into bodies of water. Even though point source pollution only originates at one location, it can have an impact on miles of streams and the ocean. The most common places for PS to be found are in municipal sewage discharges from densely populated areas or areas with a high concentration of homes, as well as in industrial wastewater loads from various businesses. The introduction of systematic regulations, standards, and relatively high-quality engineering practices in the last many years has been largely responsible for controlled PS.

10.2.2 Non-point Source

The numerous diffuse causes that contribute to non-point source (NPS) pollution are different from those that originate in industrial and sewage treatment plants. The term "NPS pollution" refers to the contamination of waterways like as rivers, lakes, and coastlines by pollutants carried by runoff from precipitation, snowmelt, or irrigation water. Concern among the general population over NPS contamination from farms has been on the rise as a leading indicator of declining surface water quality. The movement of precipitation or snowmelt over and through the earth produces NPS pollution. Runoff collects contaminants from the environment and human activities on its way to various bodies of water, including rivers, wetlands, coastal waterways, and groundwater systems. Hence, nonpoint source pollution is the name given to pollution originating from scattered sources. These could include rainfall overflow, agricultural runoff, or rubbish blown into streams from the surrounding area. The main source of water pollution in several countries in waters is nonpoint source contamination, but because there isn't just one offender, it can be challenging to control. The characteristics of non-point source (NPS) pollution differ from those of point source pollution, including the fact that it occurs at random intervals, is caused by complicated mechanisms and processes, has unknown discharge pathways and volumes, varies in pollutant loads throughout space and time, and is difficult to monitor, simulate, and regulate. Because of these features, NPS pollution is difficult to detect, regulate, and cure (Shen et al. 2012). The East River's PS and NPS pollution processes were examined using the SWAT, a multi-disciplinary modelling technique by Wu and Chen (2013). They took four major objectives to this research: (1) determine the current state of the river's water quality at two significant points in its course; (2) study how nutrient levels change with the seasons as a result of hydrometeorological forcing; (3) pinpoint the locations where NPS nutrient loadings are most concentrated; and (4) analyse the watershed-scale effects of PS and NPS pollutant loadings.

10.2.3 Transboundary

It should be clear that a line on a map cannot regulate water contamination. Water contamination that originates in one nation and extends to another is referred to as transboundary pollution. An area can get contaminated by both natural disasters such as oil spills and the gradual seepage of industrial, agricultural, or municipal trash downstream. Types of water being impacted can be deliberated are discussed further.

Groundwater Pollution

One of our most valuable but often overlooked natural resources is groundwater, which forms when precipitation seeps into the soil and fills aquifers (basically underground reservoirs of water) through their cracks, fissures, and pores. Pumped groundwater supplies potable water to around 40% of the United States' population. When it comes to freshwater, it's the only option for some people living in rural locations. Contaminants pollute groundwater when they seep into an aquifer and render it unfit for human use. This includes anything from pesticides and fertilisers to wastewater from septic systems and landfills. Toxin removal from groundwater may be a daunting, time-consuming, and costly process. Contaminated aquifers may remain unusable for hundreds, if not thousands, of years. Contaminants may be dispersed far from their original source as groundwater seeps into bodies of water including streams, lakes, and oceans.

Surface Water Pollution

Surface water, which makes up around 70% of the planet, is responsible for the blue colour of the world's rivers, lakes, and oceans. Over 60% of the water that American homes get comes from surface water sources that are considered freshwater, meaning they are not oceanic. There is some clean water, but there is also a lot of water that isn't. According to the most recent state-wide water quality tests undertaken by the US Environmental Protection Agency, over half of our rivers and streams are contaminated, rendering them unsuitable for swimming, fishing, or drinking. Nearly one-third of our lakes fall into this category. Phosphates and nitrates, which are considered nutritional contaminants, are the most prevalent kind of pollution in these water sources. Fertiliser runoff and agricultural waste have transformed these nutrients—vital for plant and animal growth—into a major pollutant. A large quantity of contaminants are also released by industrial and municipal waste. Then there is the litter that people and businesses leave in waterways.

- **Ocean water pollution**

Eighty percent of marine pollution, also known as ocean pollution, originates on land, whether it's well inland or near the coast. Urban areas, farms, and enterprises release pollutants into our bays and estuaries, which are then carried to the ocean by streams and rivers. These contaminants include heavy metals, chemicals, pesticides, and fertilisers. Marine trash, particularly plastic, is carried by the wind together with water from sewage systems and storm drains. There is a constant flow of atmospheric carbon pollution into our rivers, and from time to time, there are both large and little oil spills and leaks that poison them. The ocean takes in as much as 25% of the carbon dioxide that people release into the atmosphere. The bulk of plastic trash is created inland; therefore, studying how rivers carry it to marine habitats may help identify and control its primary origins, as seas have been exploited as trash dumps for years despite worldwide attempts to stop this. Recently, there has been a heightened attention on the need to adopt legislation and management solutions to address the increasing production and build-up of plastic pollution in the environment.

10.3 Life Cycle Assessment (LCA)

Life cycle assessment (LCA) can answer questions about sustainable development and green terminology that no other assessment instrument can. LCA is a method for grouping the pertinent life cycle components. It is a tool for performance monitoring and possibly for better product design. LCA is not a "precise" science with ideas that can be verified, nor is it a panacea for all of our environmental issues. In other words, it's a way to figure out how a process or product will affect society, the environment, and the bottom line (Hellweg and Milà i Canals 2014). By compiling trade-offs across many environmental concerns and by evaluating the environmental sustainability of new technologies and processes, it might make a big difference. Strategic planning in the water sector may greatly benefit from LCA's quantitative approach to environmental evaluation, making it an excellent decision-support tool for assessing future operating scenarios. It can be challenging to use life cycle assessment (LCA) to explain the environmental consequences of WWTP modifications due to the complex interdependencies between the various processes that comprise WWTPs. This is because effluent from one process is often influent for another, and the impact of these interdependencies on resource consumption (e.g., energy, chemicals) is non-linear. More than just meeting conventional effluent quality standards (for things like BOD5, TSS, etc.), modern WWTPs aim to recover resources, control energy use, eliminate new contaminants, and more as part of a paradigm change. Developing a set of best practices that are relevant to this field of research can help address the complications of applying LCA to wastewater management. The usage of LCA to the wastewater sector has garnered a lot of attention over the past two decades. However, academics and practitioners lack a unified body of work that systematically explains how to conduct life cycle assessments (LCAs) for wastewater treatment plants (WWTPs), wastewater infrastructure, and wastewater management choices in general. Only broad recommendations for carrying out LCA studies are provided by ISO standards (ISO 14040 2006) and (ISO 14044 2006). Wastewater collection and treatment services had a Product Category Rules document that was released in 2014 and then expired in September 2016. The Produce Category Rules paper is helpful as a jumping off point, but it doesn't address all the challenges that can arise when applying LCA to the complicated and varied wastewater treatment industry (Corominas et al. 2020).

Protecting the environment is of the utmost importance, and LCA has shown to be an excellent tool for evaluating the environmental effects of wastewater treatment plants (WWTPs) both during the design and operation phases. Various literature reviews have been produced that discuss WWTPs and LCA research. In an effort to determine what has been accomplished and to portray the obstacles for the future, they have conducted a thorough evaluation of LCA research on wastewater treatment, following the stages of LCA methodology. Reviewing LCA applications for WWTPs has been a common practice among researchers who have compiled summaries of LCA's water industry usage. But they only care about one metric, energy usage. Few have reviewed the literature on life cycle assessment (LCA) studies that have focused on the environmental implications of wastewater treatment systems, specifically the component of these studies that has dealt with toxicity. But it's worth mentioning that there isn't a single thorough evaluation that looks into the outcomes of WWTP-related LCA research. In order to offer a qualitative explanation of the linked environmental effect categories, Zang et al. (2015) examines pertinent WWTP-related LCA research. All wastewater treatment plants (WWTPs) that use activated sludge technology or a hybrid of the two were considered in this study. They used a method that splits WWTP systems throughout all the literature reviews into two categories: base system and extended system, in order to make it easier to understand where different environmental consequences may be coming from. The inclusion of sludge disposal in the base system is a result of the fact that sludge's properties are linked to wastewater treatment and, as a result, impact the system's environmental performances. There are three main processes that make up a wastewater treatment plant (WWTP) system, as outlined in the Product Category Rules (PCR). Wastewater collection and transportation to the treatment facility are part of the upstream operations. Building the plant, running it (including using energy for treatment procedures, making chemicals and maintenance materials, using resources, and dealing with emissions and waste from the process, among other things), and finally dismantling it are all essential operations. For example, biogas may provide energy and heat, and residuals can be recycled as part of downstream operations. Wastewater treatment plants (WWTPs) typically take wastewater, energy, and chemicals as inputs into their operational stages, and release a variety of pollutants into the environment as outputs. These include water (mostly in the form of wastewater effluent), air (CO_2, CH_4, etc.), and soil (mostly as a result of agricultural applications, landfill sludge, treated wastewater irrigation, etc.). Their focus (Zang et al. 2015) was on the downstream and core process operating phases, rather than the upstream and core process building and teardown phases.

10.4 Multiple Approaches to LCA

A product or service's possible environmental consequences can be evaluated over the course of its life cycle and in connection to its function through life cycle assessment (LCA).

The Life Cycle Assessment (LCA) is most commonly used to assess a WWTP's environmental impact. When analysing a WWTP, the LCA takes into account resource and energy use, air emissions, waste generation, and process efficiency, as well as ultimate effluent quality. The life cycle assessment (LCA) assesses the whole lifespan of a wastewater treatment plant (WWTP), not only the impact of effluent discharge on water bodies, and may assist in process selection and decision-making with regard to environmental consequences. Based on the method, starting and end processes, and elements involved, there are different approaches to life cycle assessment (LCA), as discussed below and shown in Fig. 10.1.

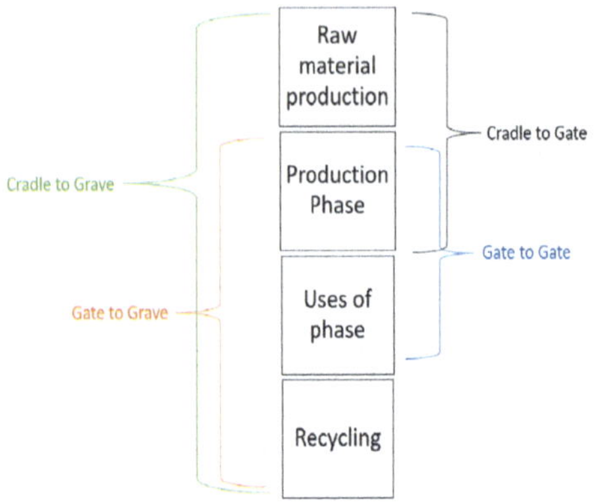

Fig. 10.1 Different LCA methods

10.4.1 Cradle-to-Grave

From the initial procurement of raw materials all the way to its eventual disposal, the term "cradle-to-grave" covers every step of a product's existence. Life cycle evaluation is one way to measure these effects, both good and bad. It takes into account the whole environmental effect of a product, from the extraction of raw materials to its manufacture, use, and eventual disposal. It begins in Fig. 10.2 with the mining of raw materials for the product and ends when the earth receives back all of its resources. Life cycle assessment (LCA) is a crucial analytical technique for determining the environmental impacts of a product from the beginning of its production all the way to its final disposal. This includes not only the extraction of raw materials but also their transportation and final disposal.

10.4.2 Cradle-to-Gate

A "cradle-to-gate" analysis looks at the whole product life cycle, beginning with the extraction of raw materials and ending at the factory gate. Business-to-business environmental product declarations are occasionally based on this assessment. A life cycle assessment (LCA) must evaluate the environmental implications of various technology alternatives or processes, and this can only be done with a well-defined system, functional units, and comparable scenarios. Product systems are defined by the criteria and unit processes that are part of them, which are defined by the system boundaries. One of the difficulties of LCA is defining the limit of the system since they differ greatly. The majority of studies have ignored the built systems' eventual decommissioning as they have taken a process-oriented, cradle-to-gate approach and have focused solely on the infrastructure's building and the tertiary treatment's operating phases. The operational phase of wastewater treatment systems was the exclusive focus of about forty percent of the research. Previous research found that

Fig. 10.2 Cradle to grave approach

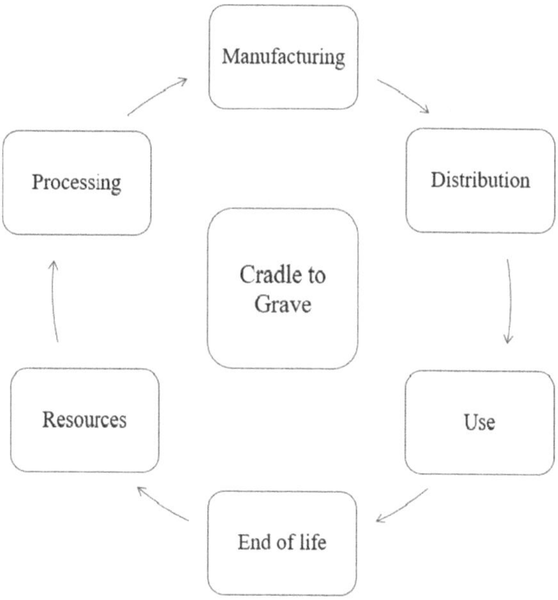

infrastructure had a low impact on overall impacts, and this is supported by the fact that the wastewater treatment plant runs regardless of whether or not its discharge is used for irrigation (Hetherington et al. 2014). Another reason given was that infrastructure's contribution to total impacts is negligible compared to the operation phase. There was some inclusion of infrastructure and the end-of-life or disposal of used consumables (such as membranes). In an LCA, incomparable outcomes may occur if the system boundaries are too narrow, failing to capture all process impacts and omitting some life cycle stages (Mehmeti and Canaj 2022).

10.4.3 Cradle-to-Cradle

While life cycle assessment (LCA) is one way to look at sustainability in design, Cradle to Cradle is another. While the word has been used by others, it was not until 2002 in the book "Cradle to Cradle—Remaking the way we make things" by Braungart and McDonough (2009) that the notion was fully articulated. Produced goods can apply for the Cradle to Cradle label at varying degrees of conformity with the concept's articulated standards, which provide the foundation for certification. Three guiding concepts form the basis of the Cradle to Cradle idea and its underlying criteria: waste is equivalent to food, use current solar income, and celebrate diversity.

A product is said to be "cradle-to-cradle" if all of its parts and materials can be recycled or used again indefinitely. By doing this, items become less harmful to the environment and become "circular," as seen in Fig. 10.3.

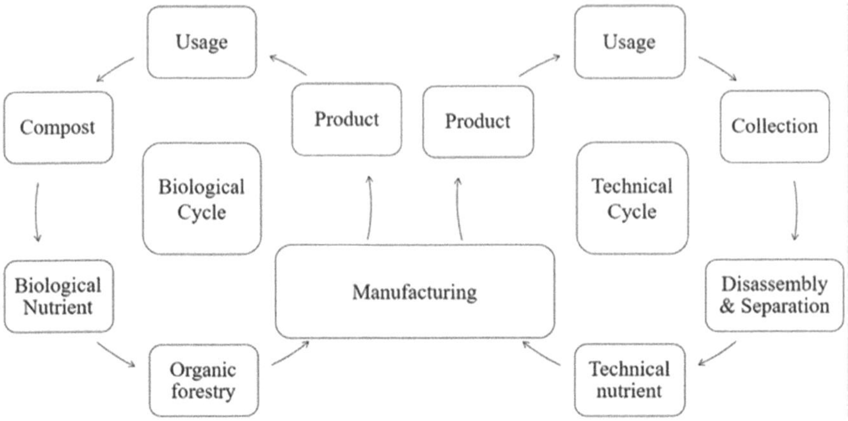

Fig. 10.3 Cradle-to-cradle approach

10.4.4 Gate-to-Gate

A gate-to-gate LCA isolates a single value-added step in the manufacturing process rather than analysing the entire chain. Once linked to the appropriate supply chain, this module may form the basis of a comprehensive cradle-to-gate evaluation. A life-cycle assessment (LCA) of a complex treatment system takes into account everything from emissions and auxiliaries to energy consumption and the overall environmental effect. An important factor preventing LCA from being widely used is the dearth of information about wastewater-specific LCIs. The purpose of this research is to provide a modular gate-to-gate inventory model for use in the chemical and related industries' wastewater treatment processes. As a result, inventory parameters may be computed according to the wastewater composition and the methods used. To achieve this goal, researchers (Köhler et al. 2007) surveyed the chemical sector for information on process parameters, effluent composition, energy and auxiliary use, and more. Hence, they were able to establish a chain of events including wastewater inflow, emissions, and technological inputs. A generic inventory model was developed based on these causal linkages.

10.4.5 Well-to-Wheel

A non-standardized technique for estimating how transportation fuels and automobiles affect energy and climate change is the well-to-wheel study (Ramachandran and Stimming 2015). In the standard lifespan of automotive technology, there are two main phases to the life cycle, and these are the fuel cycle and the vehicle cycle. Future vehicle fuel and technology selection relies heavily on fuel lifecycle analysis, sometimes called well-to-wheel analysis. Hereafter referred to as "WtW analysis," this method examines the amount of energy consumed and greenhouse gas emitted from both the fuel's manufacture and its subsequent usage in the engine or vehicle. While both methods use the same system boundaries, a WtW analysis differs from an LCA in that it disregards the energy and emissions produced when building the facilities, transporting the cars, using other materials, water, and disposing of the product at the termination of its fruitful life. In fact, full WtW cycle have two separate phases. First, there's the well-to-tank (WtT) phase, where the feedstock is recovered or produced, then the energy source is transported and stored, and finally, the fuel is converted to the vehicle's tank. Second, there's the tank-to-wheel (TtW) phase, where the vehicle uses the fuel for travel throughout its lifetime (Khan et al. 2019).

10.5 Life Cycle Assessment of Wastewater Technologies

In this new technological era, everyone is more attracted to sustainable development and evaluation but not at the cost of the environment. The secret to guaranteeing sustainable development is LCA. The primary entities accountable for the discharge of wastewater into aquatic environments and the surrounding ecosystems are the wastewater treatment plants, or WWTPs. The focus of LCA is wastewater treatment including energy consumption, resource consumption, and environmental aspects. Recently another point such as global warming potential including the emission of CH_4 and N_2O from wastewater treatment is also in the limelight. Also, take care of micropollutants in effluent and sludge in the toxicity impact categories. During the planning, designing, upgrading, and operating phases of a wastewater treatment plant's life, it should be able to provide appropriate answers for both desired emission volumes and recovered resources. It has also become common practise to use LCA in wastewater treatment to compare the relative sustainability of alternative treatment methods and identify trade-offs. LCA of several wastewater treatment (WWT) technologies is covered in more detail.

10.5.1 Life Cycle Assessment of Sewage Sludge

The wastewater treatment plant's international image is affected by the final disposal technique of sewage sludge, an inevitable result of wastewater treatment. In March 2021, the Central Pollution Control Board (CPCB) published a report estimating that the country's urban regions will generate 72,368 million litres (MLD) of sewage daily. The rise in sludge formation and the subsequent increase in wastewater treatment are the causes of this. Sludge is being more and more viewed as a resource as opposed to a waste product since it can be converted into nutrients and its carbon may be extracted and utilised again in farming. Sludge application in agriculture reduces resource depletion by replacing the manufacture and use of mineral fertilisers from a LCA standpoint. Wastewater treatment plants face planning and budgetary hurdles when it comes to sewage sludge management. Stabilised sewage sludge is widely used as a fertiliser and soil conditioner, despite the fact that it contains pathogens, heavy metals, and trace organic contaminants, which are often accumulated in the sludge throughout wastewater treatment operations. To guarantee the safety of the public, sewage sludge must be stabilised and monitored regularly before it may be applied to land. In general, biosolids, or properly stabilised sludge, are considered safe for use on land. Anaerobic digestion and pyrolysis are two examples of stabilisation procedures that may also be employed for energy recovery. This has led to sewage sludge management being more multi-focused and governed by several laws, some of which have purposes that conflict with one other. As an alternative, anaerobic digestion or incineration can be employed to recover the energy necessary to remediate the sludge. Wastewater treatment facilities may use the energy produced by biogas internally, lowering their need for energy from outside sources. Houillon and Jolliet (2005) evaluated several methods for treating wastewater sludge, taking energy and GHG emissions into account. These methods included spreading agriculture, fluidized bed incineration, moist oxidation, pyrolysis, and incineration in cement kilns and landfills. Anaerobic digestion, pyrolysis, and sludge incineration are all energy-related processes. When it comes to eutrophication, the advantages of applying digested sludge to the land instead of fertilisers are absent from the other two choices because no nutrients are recovered. Nevertheless, there are drawbacks to land application as well, since heating procedures significantly lessen the harmful effects related to heavy metal discharge.

Thermochemical and biological pathways are used to convert biomass into bioenergy. Out of all methods of thermochemical conversion, pyrolysis is easy to operate and hence it has gained popularity. The impact of pyrolysis on the environment is assessed using the life cycle assessment (LCA) tool. The use of life cycle assessment (LCA) to manage sewage sludge has been on the rise recently as a means to decipher the knotty nature of these policies and plans. LCA provides a framework for calculating the possible environmental effects of specific systems and using that information to make educated decisions. The field of solid waste management, particularly sewage sludge treatment, has made substantial use of LCA. Nevertheless, technology modelling is frequently disregarded, even though prior research shows that assumptions, including the selection of environmental emissions and how to measure them, significantly impact LCA outcomes (Yoshida et al. 2013). Last but not least, further study is required to determine the social and economic consequences of various management strategies, as well as the long-term harmful effects on ecosystems, the possibility of carbon sequestration, energy recovery, and resource recovery.

10.5.2 Life Cycle Assessment of Municipal (Biological) Wastewater Treatment Technologies

Municipal wastewater is another large component of treated wastewater. These days, wastewater from cities is being touted as the next great renewable energy source. All of the collected municipal wastewater is delivered to a centralised treatment facility, where it undergoes a number of cleaning and purification procedures. The waste streams produced by this process treatment plant include scum, centrate, biosolids, and other liquid waste streams. Biosolids, primarily derived from excess activated sludge in aeration tanks and sludge collected at the bottom of primary and secondary clarifiers, are a significant waste stream. These biosolids are dewatered until their water content reaches 70% and then sent for final disposal, often in landfills. Because biosolids, when burned, can release toxic pollutants, heat, and energy, this approach does, however, bring up serious waste management challenges. Polychlorinated biphenyls (PCBs) are harmful to fish, wildlife, plants, and animals and can be released during the combustion of biosolids. Biosolids include heavy metals, persistent organic pollutants (POPs), and a high concentration of pathogens, all of which can harm plants and build up on soil surfaces; this makes their application as fertiliser problematic. These contaminants have the potential to biomagnify once they reach the food chain. Biodiesel and other environmentally friendly transportation fuels made from biosolids are the focus of current research into green transportation fuel conversion methods including pyrolysis, gasification, and liquefaction. The conversion of biosolids into biogas for the generation of power and heat is another typical use of anaerobic digestion. Although this approach is eco-friendly as per Life Cycle Assessment (LCA), there is still cause for concern over the discharge of heavy metals from sludge, which can harm both humans and ecosystems. During wastewater treatment, scum, a layer that forms at the top of the primary settling tank, collects a variety of contaminants, including cooking oil, animal fats, food waste, soaps, and waxes. Even though activated sludge is frequently applied to it during the same anaerobic digestion process, the thick layer it forms can hinder the digester's efficiency. Scum is treated as waste by certain establishments, which either burn it or dispose of it in landfills, raising environmental issues akin to those pertaining to biosolids. Nevertheless, research has shown that dried scum, which is composed of roughly 14% organic biosolids and 62% lipids and oils, can be utilised as a substitute energy source. By removing the oil, the University of Minnesota has effectively transformed scum into fuel-grade biodiesel. Acid hydrolysis and glycerolysis are two techniques that have been researched for extracting oil from wastewater scum as a pretreatment step before producing biodiesel. These techniques efficiently lower the amount of free fatty acids in scums, and using catalysts based on zinc greatly speeds up the reaction.

The technology proposes a 7-step energy conversion pathway as shown in Fig. 10.4.

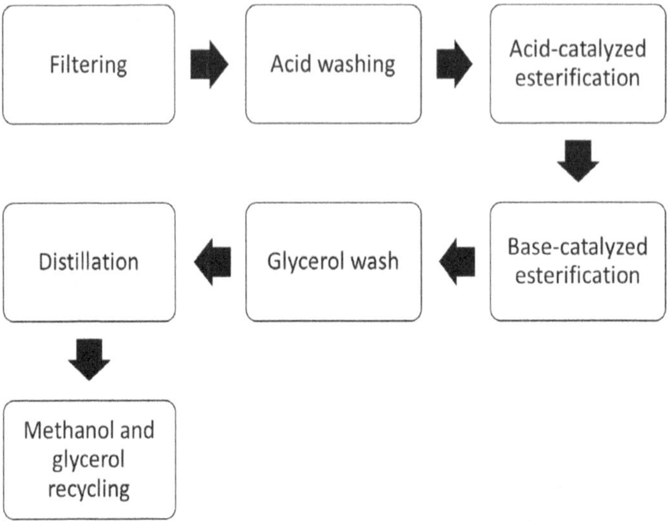

Fig. 10.4 Steps for energy conversion

The efficiency of energy conversion using scum to biodiesel technology was found to be 60% when compared to conventional scum treatment methods. The LCA determined that the scum-to-biodiesel technology posed less of a threat to eutrophication, greenhouse gas emissions, and fossil fuel depletion when compared to burning and digestion. Compared to scum digestion and combustion procedures, the scum to Biodiesel technology can yield substantial benefits and higher revenue overall. Centrate, a byproduct of the wastewater treatment process, contains high levels of pollutants and is typically recycled back into the system for further treatment. This recycling increases the pollutant levels in the aeration tank, reducing its treatment effectiveness and increasing energy use due to the need for longer aeration times. However, studies have shown that using Centrate to grow algae can result in higher yields and fewer inputs compared to other wastewater streams. Other liquid waste streams with high levels of pollutants are also typically recycled, but could be used in algae cultivation to reduce pollutants and co-produce biofuels (Naushad 2018). The pyrolysis of biomass produces biochar, a carbon-rich substance that has drawn interest as a potent wastewater adsorbent and treatment agent. Biochar produced from biomass is a promising substance that has several uses, one of which is the remediation of polluted water. Biochar is a useful resource that may be utilised in wastewater treatment to convert organic waste, such as agricultural leftovers or organic sludge, in line with the circular economy concept. When using biochar as an adsorbent in wastewater treatment processes, pollutants are removed while also capturing and retaining valuable nutrients and organic matter that can be recycled and used in agriculture or energy production. However, in order to achieve eco-sustainability, it is crucial to address the issue of secondary pollution that is expected to arise from the disposal of saturated adsorbents in the process of producing value-added materials.

10.5.3 Life Cycle Assessment of Industrial Waste Water

- Goal and scope definition;
- Life cycle inventory (LCI);
- Life cycle impact assessment (LCIA); and
- Outcome interpretation.

Above mentioned four points comprises the iterative four-phase Life Cycle Assessment (LCA) technique. The LCA framework defined by ISO 14040:2006 and ISO 14044:2006 is followed in this study (Hijrah et al. 2023). To begin, we must pinpoint the specific areas of wastewater treatment plant systems that are most responsible for the possible negative effects on the environment. Two, with and without energy recovery programmes, we need to assess the possible effects of industrial wastewater treatment systems on the environment. Determining the study's purpose, objectives, functional unit, and system boundary is the first step in conducting an LCA. To rephrase, in order to do an apples-to-apples comparison, the functional equivalent is a collection of design criteria that both samples should have. It should be mentioned that a partial life cycle assessment (LCA) focuses on a subset of a product's phases and activities within those phases, depending on the aims and parameters of the study. In the second step, known as inventory analysis or life cycle inventory (LCI), all the necessary data is gathered and the inputs and outputs are calculated using the LCI database. Based on the findings of the inventory analysis, the next step is to conduct a life cycle impact assessment (LCIA) to identify the environmental consequences across the stages of the life cycle that were established in Stage I. Possible environmental impact assessment metrics, such GWP, are generated at the LCIA stage. Interpretation of LCI and LCIA data is the last step in an LCA research (Kamali et al. 2019). Collecting the required data and producing generalizable findings is not a simple feat; LCA is a complex procedure. Life cycle assessment (LCA) is not a science of precise measurement but rather a science of best guesses due to the uncertainties inherent in the process. Simply said, LCA is a useful tool for comparing the relative performance of different possibilities, which in turn helps to narrow down the course of future decisions.

Goal and Scope

This functional unit is selected in accordance with the study's objectives; it specifies the function of the system. To be able to compare systems, their functions, or functional units, must be identical, as must their boundaries and assumptions. There are a lot of factors to think about while treating water, such as the reference water flow, pollution costs, and the plant's treatment efficiency (Suh and Rousseaux 2002). Additionally, the boundaries of both space and time are defined at this very moment. It is possible to measure the impact of a system across its lifespan. Along with planning, construction, operation, and destruction, the life cycle of a sewage treatment plant begins with excavation. Energy production, transportation, sludge removal, and the production of building materials and reagents are all part of the incidental operations (Renou et al. 2008). Each product system goes through three distinct phases: construction (the materials needed to create the plant), operations (the materials and energy spent during plant operation), and destruction or end-of-life (waste disposal). Since the operational stage contributes significantly to the environmental load, it is important to take the gate-to-gate phase into consideration. Even so, the environmental impact during construction was not as great. The fact that the setup of individual reactor is composed of different materials makes it difficult to determine the emissions from the construction phase, lending credence to the argument that this phase was ignored. Environmental damage was mostly caused by the reactor's heating and the addition of the chemical NaOH during the operating phase, as was shown in the previous case. All assessment phases were considered and confirmed this. Considering the gate-to-gate phase is crucial since the operational stage is a major contributor to the environmental burden. Because of the high stakes involved in managing a waste water treatment plant—which might have far-reaching consequences for the environment if not addressed—this study zeroes emphasis on that part of the process (Hijrah et al. 2023).

Life Cycle Inventory

The second LCA step, inventory analysis, covers data gathering and measurement methods for every input and output of a product system. Within the system's limits and the designated functional unit, this comprises energy, resources, transportation, goods, coproducts, and waste discharge to air, water, and soil. From the estimated mass and energy balances of the system in question, a list of input and output flows is generated using the data gathered in this section. Particularly when dealing with missing data, it is important to note any assumptions made at that phase. The time it takes to gather data for an LCA is substantial. As part of a LCA, a Life Cycle Inventory (LCI) compiles and quantifies the inputs and outputs for a specific product system over the course of its lifetime. Cleaner manufacturing using LCI has been around since the 1960s, and the idea has gained a lot of traction in academia and industry in the previous several decades. With a few exceptions, LCI has been thought of as a relatively simple process when compared to the other parts of LCA. But contrary to popular assumption, LCI may be calculated using a variety of ways, and these approaches frequently provide quite different outcomes. Hence, it's important to weigh the pros and cons of various LCI approaches and choose the right one (or ones) for the job (Suh and Huppes 2005).

In three easy stages, the life cycle inventory may be explained.

(1) Information on the general operations or technology used in wastewater treatment plants, obtained by visits to the facilities or communications with wastewater treatment facility officials
(2) Data collecting, which takes into account the kind, source, and needs for the data

(3) Data evaluation and resolution the rigorous processes used to detect and isolate faults.

Environmental impact assessments (ECAs) of WWTPs and other end-of-pipe technologies seek to quantify environmental emissions from all processes, compile them in an LCI, and then transform them into indicators and effects on the environment (Yoshida et al. 2013). Important steps in conducting an LCA research, such as data collecting and LCI build-up, are often constrained by the accessibility of trustworthy data. An essential piece of information that aids in data collecting is that wastewater treatment plants already track water and sludge quality indicators as part of mandated laws, which they submit to authorities on a regular basis. Periodic reports on influent and effluent composition only provide normal operating characteristics; the evaluation of greenhouse gas emissions or PPCPs is not included. Emerging contaminants in sewage are another modern-day worry. When comparing this advanced LCI (which includes GHG and PPCP emissions) to a less comprehensive method based on traditional characteristics, the study done on (Lorenzo-Toja et al. 2016) overarching goal was to show how much better this latter method is for environmental evaluation of WWTPs. Rather than depending on assumed literature values, this effort by the researchers aimed to provide more accurate emission factors for the compounds stated before. Consequently, the study's findings might help establish new standards for the ecological efficiency of WWTPs.

Life Cycle Impact (LCIA) and Interpretation

Three parts make up LCA standards; that one is the life cycle impact assessment (LCIA). In this step, the LCI data is converted into environmental effect ratings. These ratings, together with the LCI data, are then examined in the interpretation stage (Generowicz et al. 2023). LCIA includes both required and optional components. It is necessary to choose effect categories, classify inventory data into relevant impact categories, and characterize inventory data within impact categories. Normalization, grouping, weighing, and data quality analysis are optional components.

Human health, ecosystems, and resource shortage were the three endpoints that were considered (Hijrah et al. 2023). Ecosystems depict the number of species lost over time in a particular location, but disability-adjusted life years or years of life lost were employed to highlight how medical interventions affected people's health (DALY). Conversely, the excess cost of future resource output over an infinite period of time is described as resource scarcity.

Endpoint analysis findings show that the techniques used to treat wastewater seem to have the greatest impact on human health. Impact potentials are primarily contributed by three factors: hazardous waste, polyaluminium chloride, PAC, and power are the first three. Hazardous waste (sludge) accounts for around 53.4% of the possible effects on human health. In line with this, research conducted by the Economic Planning Unit (EPU) revealed that planned waste has been progressively increasing from around three million tonnes in 2015 to four million tonnes in 2019.

Accordingly, the circular economy model's emphasis on the life cycle perspective (recovery, reuse, reduce, recycle) provides one strategy for impact management and influence. This showed that hazardous waste with a heat recovery system was less likely to negatively affect the environment, with the likelihood decreasing from 53.4% (WWTP without HRS) to 41.2%. In place of the current conventional fossil fuel power supply, the facility generates electricity using the steam produced by the HRS.

Additionally, chemical use, such as the use of polyaluminium chloride (PAC), is shown to be one of the biggest contributors to the negative effects on the environment (38.3%), second only to hazardous waste. According to Sabarwal et al. (2018), excessive chemical use may have a negative influence on the health and well-being of humankind if the eradication plan is not maintained well. Additionally, PAC is a caustic chemical, thus it is crucial to take precautions while handling the material, per the PAC material safety data sheet.

The danger to human health from enormous chemical input intake may be increased. But with the introduction of HRS, as was already mentioned, the results revealed a considerable 12.2% reduction in the effect of hazardous waste. Consequently, PAC is now the leading hazardous waste in the HRS system, accounting for 47.4% of all impact potentials. Be aware that the effect reduction endeavour in this research primarily focuses on reducing hazardous trash. The effect reduction research on PAC needs further inventory work.

Additionally, power is being used for plant operations, according to the electricity mix in Malaysia. According to the Malaysia Energy Statistics Handbook 2020 (Yim and Pok 2020), Malaysia's primary sources of energy are still nonrenewable sources, primarily coal (42.8%) and gas (40.2%). As a result, burning fossil fuels for electricity generation will have an adverse effect on these resources' long-term availability and emit greenhouse gases like carbon dioxide (CO_2) and nitrous oxide (N_2O) into the environment. This combustion also releases particulates, which degrade the quality of the surrounding air. These gaseous substances can have lasting effects on human health if ingested. Since the possible effects of energy on the environment are very substantial air. These gaseous substances can have lasting effects on humankind if ingested. Wastewater treatment facilities' environmental effect capacity are greatly impacted by electricity, thus further inventory work is necessary to evaluate this effect, especially for the plant with HRS.

A minimum of two LCIA methodologies, such as sensitivity analysis, should be used in an LCA research to ensure that the choices made had a significant impact on the outcomes.

There are two major ways to determine the findings of a life cycle impact assessment (LCIA): midway and endpoint. Before reaching the endpoints, where characterisation factors or indicators may be developed to indicate the relative importance of emissions or extractions, midpoints are seen as links in the cause-effect chain (an environmental mechanism) of an impact category. The propensity for photochemical ozone (smog) formation, acidification, eutrophication, ozone depletion, and global warming are all instances of common midway characterisation elements (Mehmeti and Canaj 2022). However, closer to the end of the chain are the endpoint indicators, which deal with the real harm that these compounds may do when eaten or released (such as harm to people's health, the environment, and resources). Product environmental profiles are more thoroughly described when a larger number of effect categories are included. Midpoint and endpoint approaches, when combined, may yield valuable insights for many stakeholders. More frequent and extensive reporting of uncertainty analysis is needed as sensitivity analysis in combination with it is lacking in the present research (Mehmeti and Canaj 2022).

Polymer Members for Water Treatment

Polymeric materials have gained significant attention in water treatment applications due to their versatility, cost-effectiveness, and ability to be tailored for specific purposes. These materials can be broadly categorized into synthetic polymers and natural polymers, each offering unique advantages in water purification processes.

Synthetic polymers, such as polyethersulfone (PES), polyvinylidene fluoride (PVDF), and polyacrylonitrile (PAN), are widely used in membrane-based water treatment systems. These polymers exhibit excellent chemical and thermal stability, making them suitable for various separation processes, including ultrafiltration, nanofiltration, and reverse osmosis.

PES, for instance, has been extensively studied for its high performance in water treatment applications. Its hydrophilicity and good mechanical properties make it an ideal candidate for membrane fabrication. A study demonstrated the effectiveness of PES membranes in removing heavy metals from wastewater, achieving removal efficiencies of over 90% for lead and cadmium ions (López Días et al. 2019).

PVDF membranes have also shown promise in water treatment applications, particularly due to their excellent chemical resistance and mechanical strength. A study explored the use of PVDF membranes modified with graphene oxide for the removal of organic pollutants from water. The modified membranes exhibited enhanced permeability and antifouling properties compared to pristine PVDF membranes (Pardo et al. 2021).

Natural polymers, such as chitosan and alginate, have also gained attention in water treatment applications due to their biodegradability and abundance. Chitosan, derived from chitin found in crustacean shells, has shown excellent adsorption capabilities for various pollutants, including heavy metals and dyes. A comprehensive review by Vakili et al. (2019) highlighted the potential of chitosan-based materials in water and wastewater treatment, emphasizing their eco-friendly nature and versatility (Vakili et al. 2019).

Polymer Composites for Seawater Treatment

Seawater treatment presents unique challenges due to the high salinity and presence of various contaminants. Polymer composites have emerged as promising materials for addressing these challenges, offering improved performance and durability compared to traditional polymeric materials.

One of the most significant applications of polymer composites in seawater treatment is in the development of advanced reverse osmosis (RO) membranes. These membranes typically consist of a thin film composite (TFC) structure, with a polyamide active layer supported by a porous substrate. Recent research has focused on incorporating nanomaterials into these composites to enhance their performance and fouling resistance.

A study by Yin et al. (2016) investigated the use of carbon nanotube (CNT)-incorporated polyamide TFC membranes for seawater desalination. The researchers found that the addition of CNTs improved the membrane's water permeability and salt rejection capabilities, while also enhancing its resistance to biofouling (Yin et al. 2016).

Another promising approach in polymer composites for seawater treatment involves the use of nanocomposite membranes. These membranes incorporate nanoparticles or nanomaterials into the polymer matrix to enhance specific properties. For instance, Yin et al. (2016) developed a graphene oxide (GO)-enhanced polyamide thin-film nanocomposite membrane for water purification. The inclusion of GO improved the membrane's water flux and salt rejection performance, making it suitable for seawater desalination applications (Yin et al. 2016).

Polymer composites have also been explored for the removal of specific contaminants from seawater. For example, a novel magnetic nanocomposite based on chitosan and iron oxide nanoparticles for the elimination of heavy metals from seawater. The composite exhibited high adsorption capacities for lead and cadmium ions, demonstrating its potential for treating metal-contaminated seawater (Abdel-Moneim et al. 2020).

Cellulose Membranes for Water Treatment

Cellulose, as the most abundant natural polymer on Earth, has gained significant attention in water treatment applications due to its biodegradability, renewability, and versatile chemical structure. Cellulose-based membranes offer several advantages, including high hydrophilicity, good mechanical strength, and the potential for easy surface modification.

One of the most promising forms of cellulose for membrane fabrication is nanocellulose, which includes cellulose nanofibrils (CNF) and cellulose nanocrystals (CNC). These nanomaterials exhibit exceptional mechanical properties and high surface area, making them ideal for water filtration applications. A comprehensive review by Mahfoudhi and Boufi (2017) highlighted the potential of nanocellulose-based materials in water treatment, discussing their preparation methods, properties, and applications in various purification processes (Mahfoudhi and Boufi 2017).

Recent research has focused on developing composite membranes that combine cellulose with other materials to enhance their performance. For instance, cellulose acetate/graphene oxide composite membrane for water desalination. The incorporation of graphene oxide improved the membrane's water flux and salt rejection capabilities, demonstrating its potential for seawater desalination applications (Bacchin 2018).

Another innovative approach in cellulose-based membranes involves the use of bacterial cellulose (BC). BC is produced by certain bacteria and possesses unique properties, including high purity, crystallinity, and mechanical strength. BC-based membrane functionalized with polyethyleneimine can be used for the removal of heavy metal ions from water. The membrane exhibited excellent adsorption capacities for various metal ions, including copper, lead, and cadmium (Kwon et al. 2019).

Biocomposites for Water Treatment

Biocomposites, which combine natural polymers with other materials, have emerged as promising candidates for water treatment applications due to their eco-friendly nature and unique properties. These materials often exhibit enhanced performance compared to their individual components, making them suitable for various water purification processes.

One of the most studied biocomposites for water treatment is chitosan-based materials. Chitosan, derived from chitin, offers excellent adsorption properties and can be easily modified to enhance its performance. For example, a chitosan-based biocomposite by cross-linking chitosan with epichlorohydrin and grafting it with ethylenediamine. The resulting material showed improved adsorption capacities for copper ions compared to unmodified chitosan (Wan Ngah et al. 2011).

Alginate-based biocomposites have also shown promise in water treatment applications. Alginate, a polysaccharide derived from brown seaweed, can form hydrogels with excellent water absorption properties. Gan et al. (2018) developed a novel alginate/graphene oxide biocomposite hydrogel for the removal of methylene blue dye from water. The biocomposite exhibited high adsorption capacity and good reusability, demonstrating its potential for practical water treatment applications (Yu et al. 2018).

Cellulose-based biocomposites have also gained attention in water treatment research. Mohamed et al. developed a biocomposite membrane by combining cellulose nanofibers with polyvinyl alcohol (PVA) and graphene oxide. The resulting membrane showed excellent performance in removing various water pollutants, including heavy metals and organic dyes (Mohamed et al. 2015).

In conclusion, polymer members, polymer composites, cellulose membranes, and biocomposites offer a wide range of solutions for water treatment applications. These materials have demonstrated excellent performance in various aspects of water purification, including desalination, heavy metal removal, and organic pollutant adsorption. Ongoing research continues to explore novel combinations and modifications of these materials to address the growing challenges in water treatment and environmental remediation.

As the field of water treatment materials continues to evolve, future research directions may include:

1. Development of multifunctional biocomposites that can simultaneously address multiple water treatment challenges.
2. Exploration of novel natural polymers and their derivatives for sustainable water treatment solutions.
3. Integration of advanced nanomaterials with biopolymers to create high-performance water treatment membranes and adsorbents.

4. Investigation of scalable and cost-effective production methods for these advanced materials to facilitate their widespread adoption in water treatment applications.

By leveraging the unique properties of polymers, composites, and biocomposites, researchers can develop innovative solutions to address the global challenges of water scarcity and contamination, ultimately contributing to improved water quality and sustainability.

10.6 Conclusion

Life cycle assessment is a powerful tool to quantify the environmental sustainability of various products, including bioenergy production. It also can identify the sources and causes of the environmental impacts of bioenergy product systems. There are also drawbacks and restrictions to employing life cycle assessment, even with its many benefits in evaluating the environmental sustainability of bioenergy product systems. Wide variations in the final conclusions might arise from different techniques of effect assessment, different inventory data, different assumptions, and several other sources of uncertainty. These problems may have a detrimental impact on the dependability and accuracy of life cycle assessment results for bioenergy product systems, which might result in poor policy and decision-making.

By cleaning wastewater and lowering the amount of toxins released into the environment, wastewater treatment facilities significantly contribute to decreasing pollution. Even yet, their business model may leave some kind of ecological imprint. So, to measure the environmental effect of plants in many ways, Life Cycle Assessment (LCA) offers a thorough framework for wastewater treatment.

A diversity of approaches may be utilised to estimate Life Cycle Assessment (LCA), which is essential for sustainable economic growth and depends on the process, its starting stage, its ultimate stage, and its constituents. One byproduct of wastewater treatment, sewage sludge, has dual potential as both an agricultural input and an energy source. The purpose of a LCA is to compare the ecological effects of various management decisions. Centrate and other liquid waste streams typically recycled by the system include high concentrations of pollutants.

With a focus on the operational phase, this research evaluates the possible environmental implications at gate-to-gate barriers. The three areas that may have the most negative impact on the environment were power, raw material utilisation, and hazardous waste. The generation of hazardous waste dwarfs the use of enzymes, anionic resin, cationic resin, and activated carbon, among other components. In order to manage the existing surge in water demands and plan for waste formation, it is necessary to create a framework for sustainable, superior practices in managing industrial wastewater treatment systems while considering the life cycle viewpoint. One step in doing a LCA is creating an inventory of all the resources used by a product or service and how much of each they use. This process is known as a Life

Cycle Inventory (LCI). With the exception of a few problems like allocation, LCI has been thought of as a relatively simple process in comparison to the other parts of LCA. To help decision-makers find the best eco-friendly solutions, LCA studies analyse the environmental implications of various scenarios or treatment technologies. With this data, the plant can better allocate resources, run more efficiently, and reduce their impact on the environment. A minimum of two LCIA methodologies, such as sensitivity analysis, should be used in an LCA research to ensure that the choices made had a significant impact on the outcomes.

In the years to come, wastewater reuse is predicted to see massive expansion. As a result, LCA would be in high demand when it comes to strategic planning and decision-making. The water and wastewater sector has made good use of life cycle assessment (LCA) so far, but there has to be further study to determine the environmental impacts and give scientific recommendations for the sustainable use of reclaimed water on farms.

In a nutshell, when combined with other methods such as chemical and microbial risk analysis (used by local authorities to set discharge limits) and environmental impact assessment (needed for large wastewater treatment plants), conceptual, simplified and detailed LCA is utmost important to evaluate the sustainability of wastewater treatment systems.

References

Abdel-Moneim A et al (2020) Novel polydatin-loaded chitosan nanoparticles for safe and efficient type 2 diabetes therapy: in silico, in vitro and in vivo approaches. Int J Biol Macromol 154:1496–1504. https://doi.org/10.1016/j.ijbiomac.2019.11.031

Abily M et al (2023) Strategic routes for wastewater treatment plant upgrades to reduce micropollutants in European surface water bodies. J Clean Prod 415:137867. https://doi.org/10.1016/j.jclepro.2023.137867

Babuji P et al (2023) Human health risks due to exposure to water pollution: a review. Water 15(14):2532. https://doi.org/10.3390/w15142532

Bacchin P (2018) Membranes: a variety of energy landscapes for many transfer opportunities. Membranes 8(1):10. https://doi.org/10.3390/membranes8010010

Braungart M, McDonough W (2009) Cradle to cradle. Random House

Corominas L et al (2020) The application of life cycle assessment (LCA) to wastewater treatment: a best practice guide and critical review. Water Res 184:116058. https://doi.org/10.1016/j.watres.2020.116058

Gan L et al (2018) Graphene oxide incorporated alginate hydrogel beads for the removal of various organic dyes and bisphenol a in water. Colloid Polym Sci 296(3):607–615. https://doi.org/10.1007/s00396-018-4281-3

Generowicz A et al (2023) Life cycle assessment for the environmental impact assessment of a city cleaning system. The case of Cracow (Poland). J Clean Prod 382:135184. https://doi.org/10.1016/j.jclepro.2022.135184

Guven H et al (2023) Energy and material refineries of future: wastewater treatment plants. J Environ Manag 329:117130. https://doi.org/10.1016/j.jenvman.2022.117130

Hellweg S, Milà i Canals L (2014) Emerging approaches, challenges and opportunities in life cycle assessment. Science 344(6188):1109–1113. https://doi.org/10.1126/science.1248361

Hellweg S et al (2023) Life-cycle assessment to guide solutions for the triple planetary crisis. Nat Rev Earth Environ 4(7):471–486. https://doi.org/10.1038/s43017-023-00449-2

Hetherington AC et al (2014) Use of LCA as a development tool within early research: challenges and issues across different sectors. Int J Life Cycle Assess 19(1):130–143. https://doi.org/10.1007/s11367-013-0627-8

Hijrah N, Syakir MI, Syah BI (2023) Quantification of environmental impact potentials within the processes of industrial wastewater treatment system. IOP Conf Ser: Earth Environ Sci 1167(1):012019. https://doi.org/10.1088/1755-1315/1167/1/012019

Houillon G, Jolliet O (2005) Life cycle assessment of processes for the treatment of wastewater urban sludge: energy and global warming analysis. J Clean Prod 13(3):287–299. https://doi.org/10.1016/j.jclepro.2004.02.022

Kamali M, Hewage K, Sadiq R (2019) Conventional versus modular construction methods: a comparative cradle-to-gate LCA for residential buildings. Energy Build 204:109479. https://doi.org/10.1016/j.enbuild.2019.109479

Khan MI et al (2019) Life cycle (well-to-wheel) energy and environmental assessment of natural gas as transportation fuel in Pakistan. Appl Energy 242:1738–1752. https://doi.org/10.1016/j.apenergy.2019.03.196

Köhler A et al (2007) Input-dependent life-cycle inventory model of industrial wastewater-treatment processes in the chemical sector. Environ Sci Technol 41(15):5515–5522. https://doi.org/10.1021/es0617284

Kwon G et al (2019) Nanoporous cellulose paper-based SERS platform for multiplex detection of hazardous pesticides. Cellulose 26(8):4935–4944. https://doi.org/10.1007/s10570-019-02427-8

López Días V et al (2019) The use of Nafion membranes to measure $^{2}H/^{1}H$ and $^{18}O/^{16}O$ isotopic ratios in water. J Membr Sci 572:128–139. https://doi.org/10.1016/j.memsci.2018.11.003

Lorenzo-Toja Y et al (2016) Beyond the conventional life cycle inventory in wastewater treatment plants. Sci Total Environ 553:71–82. https://doi.org/10.1016/j.scitotenv.2016.02.073

Mahfoudhi N, Boufi S (2017) Nanocellulose as a novel nanostructured adsorbent for environmental remediation: a review. Cellulose 24(3):1171–1197. https://doi.org/10.1007/s10570-017-1194-0

Mehmeti A, Canaj K (2022) Environmental assessment of wastewater treatment and reuse for irrigation: a mini-review of LCA studies. Resources 11(10):94. https://doi.org/10.3390/resources11100094

Mohamed MA et al (2015) Physicochemical properties of "green" nanocrystalline cellulose isolated from recycled newspaper. RSC Adv 5(38):29842–29849. https://doi.org/10.1039/C4RA17020B

Naushad M (2018) Naushad Mu (ed) Life cycle assessment of wastewater treatment. CRC Press. https://doi.org/10.1201/9781315165820

Pardo F, Zarca G, Urtiaga A (2021) Effect of feed pressure and long-term separation performance of Pebax-ionic liquid membranes for the recovery of difluoromethane (R32) from refrigerant mixture R410A. J Membr Sci 618:118744. https://doi.org/10.1016/j.memsci.2020.118744

Ramachandran S, Stimming U (2015) Well to wheel analysis of low carbon alternatives for road traffic. Energy Environ Sci 8(11):3313–3324. https://doi.org/10.1039/C5EE01512J

Renou S et al (2008) Influence of impact assessment methods in wastewater treatment LCA. J Clean Prod 16(10):1098–1105. https://doi.org/10.1016/j.jclepro.2007.06.003

Sabarwal A, Kumar K, Singh RP (2018) Hazardous effects of chemical pesticides on human health—cancer and other associated disorders. Environ Toxicol Pharmacol 63:103–114. https://doi.org/10.1016/j.etap.2018.08.018

Shen Z et al (2012) An overview of research on agricultural non-point source pollution modelling in China. Sep Purif Technol 84:104–111. https://doi.org/10.1016/j.seppur.2011.01.018

Suh S, Huppes G (2005) Methods for life cycle inventory of a product. J Clean Prod 13(7):687–697. https://doi.org/10.1016/j.jclepro.2003.04.001

Suh Y-J, Rousseaux P (2002) An LCA of alternative wastewater sludge treatment scenarios. Resour, Conserv Recycl 35(3):191–200. https://doi.org/10.1016/S0921-3449(01)00120-3

Tsangas M et al (2023) LCA of municipal wastewater treatment. Chemosphere 341:139952. https://doi.org/10.1016/j.chemosphere.2023.139952

Vakili M et al (2019) Regeneration of chitosan-based adsorbents used in heavy metal adsorption: a review. Sep Purif Technol 224:373–387. https://doi.org/10.1016/j.seppur.2019.05.040

Wan Ngah WS, Teong LC, Hanafiah MAKM (2011) Adsorption of dyes and heavy metal ions by chitosan composites: a review. Carbohydr Polym 83(4):1446–1456. https://doi.org/10.1016/j.carbpol.2010.11.004

Wu Y, Chen J (2013) Investigating the effects of point source and nonpoint source pollution on the water quality of the East River (Dongjiang) in South China. Ecol Indic 32:294–304. https://doi.org/10.1016/j.ecolind.2013.04.002

Yim M-S, Pok PJ (2020) Examination of the possibility of nuclear-renewable hybrid energy systems in Malaysia, p 020002. https://doi.org/10.1063/5.0031558

Yin J, Zhu G, Deng B (2016) Graphene oxide (GO) enhanced polyamide (PA) thin-film nanocomposite (TFN) membrane for water purification. Desalination 379:93–101. https://doi.org/10.1016/j.desal.2015.11.001

Yoshida H, Christensen TH, Scheutz C (2013) Life cycle assessment of sewage sludge management: a review. Waste Manag Res: J Sustain Circ Econ 31(11):1083–1101. https://doi.org/10.1177/0734242X13504446

Yu B, Gele A, Wang L (2018) Iron oxide/lignin-based hollow carbon nanofibers nanocomposite as an application electrode materials for supercapacitors. Int J Biol Macromol 118:478–484. https://doi.org/10.1016/j.ijbiomac.2018.06.088

Zang Y et al (2015) Towards more accurate life cycle assessment of biological wastewater treatment plants: a review. J Clean Prod 107:676–692. https://doi.org/10.1016/j.jclepro.2015.05.060

The manufacturer's authorised representative in the EU is Springer Nature Customer Service Centre GmbH, Europaplatz 3, 69115 Heidelberg, Germany. If you have any concerns regarding our products, please contact ProductSafety@springernature.com

Printed and bound by CPI Group (UK) Ltd, Croydon, CR0 4YY
30/09/2026
02225572-0006